W0227437

Additional praise for

ON THE ART AND
CRAFT OF DOING SCIENCE

ON THE ART AND CRAFT
OF DOING SCIENCE

On the Art and Craft of Doing Science

KENNETH CATANIA

PRINCETON UNIVERSITY PRESS

PRINCETON & OXFORD

Published by Princeton University Press
41 William Street, Princeton, New Jersey 08540
99 Banbury Road, Oxford OX2 6JX

press.princeton.edu

All Rights Reserved

Library of Congress Cataloging-in-Publication Data

Names: Catania, Kenneth, 1965– author.
Title: On the art and craft of doing science / Kenneth Catania.
Description: Princeton : Princeton University Press, [2025] | Series: Skills for scholars | Includes bibliographical references and index.
Identifiers: LCCN 2024023701 (print) | LCCN 2024023702 (ebook) | ISBN 9780691249261 (paperback) | ISBN 9780691249278 (hardback) | ISBN 9780691249650 (ebook)
Subjects: LCSH: Creative ability in science. | Science and the arts. | Science—Vocational guidance. | BISAC: SCIENCE / General | SCIENCE / Philosophy & Social Aspects
Classification: LCC Q172.5.C74 C37 2025 (print) | LCC Q172.5.C74 (ebook) | DDC 501—dc23/eng/20240711
LC record available at https://lccn.loc.gov/2024023701
LC ebook record available at https://lccn.loc.gov/2024023702

British Library Cataloging-in-Publication Data is available

Editorial: Alison Kalett and Hallie Schaeffer
Production Editorial: Natalie Baan
Cover Design: Hunter Finch
Production: Lauren Reese
Publicity: Kate Farquhar-Thomson and Matthew Taylor
Copyeditor: Anne Cherry

Jacket image: Liliya Rodnikova / Stocksy

This book has been composed in Arno

10 9 8 7 6 5 4 3 2 1

To my parents

For teaching me to be curious, showing me the path,
and then cheering me all the way.

CONTENTS

INTRODUCTION

FIGURE 0.1. Don Quixote, by Charles Catania.

Honesty's the best policy.

—MIGUEL DE CERVANTES

THERE ARE TWO IMAGES that hang on the wall of my home office. On one side of the window is a painting of the HMS *Beagle* from Darwin's famous voyage, as the ship enters a cove in Tierra del Fuego. On the other side of the window is a drawing of Don Quixote riding off into the sunset (see figure 0.1, drawn by my father). The images are reminders of the two main ingredients that go into my research. The picture from Darwin's voyage is a reminder to collect data. The picture of Don Quixote, who imagined himself a knight and jousted with windmills, is a reminder to keep dreaming. Data and imagination, in proper combination, have been key to my life in science.

I never thought I'd have reason to mention these pictures; they are personal reminders of something most scientists seldom discuss or even admit—namely, that there are two sides to the process of discovery. One side of the process is the expected reliance on hypotheses, method, theory, models, statistics, logic, and the like. The other side of the process is far more mysterious and inscrutable. This is the realm of creativity, inspiration, imagination, and the often intangible source of new ideas.

If I had to categorize this book, I would say that it lies somewhere at the intersection of these two different realms of science. It is my best attempt, based on a career spent making discoveries, to distill from the mixture of these two worlds some advice about the craft of doing science. By the craft of science, I mean how to find and solve scientific puzzles while setting yourself up for the fun part—eureka moments when nature reveals one of her secrets.

I'm keenly aware that suggesting an approach to doing science is a bold thing. For most of my career such a project would have been inconceivable in the truest sense of the word. Inconceivable because my own work seemed to follow a chaotic path of exploration punctuated by luck—hardly a style from which to draw lessons. So I should tell you how I got here and why I thought I should write this book.

The seed for this project was a discovery, actually several discoveries, made not in the laboratory, but rather made when I decided to take a break from my experiments to write my last book. The book is called *Great Adaptations*, and it covers many of the studies that I have conducted during my career. These include work on star-nosed moles, shrews, electric eels, snakes, zombie-making wasps, earthworms, and even some strange traditions practiced by humans. The prospect of writing a book is daunting. Before I started writing I did some homework by reading every book of advice I could find on the subject, including William Zinsser's *On Writing Well*, Anne Lamott's *Bird by Bird*, Mary Karr's *The Art of Memoir*, and Jon Franklin's *Writing for Story*, to mention just a few.[1]

I was surprised to feel an immediate kinship with these writers and a connection to their struggles. There's the question of picking the right topic, the mystery of the process, the impossibility of predicting the plot, the idiosyncratic strategies for success, and most especially the ideas that seem to come out of nowhere—at least when things are going well. I experience all of this while doing science. Even the concept of a muse resonates as a metaphor for those special days when, for no obvious reason, the creative process kicks into overdrive.

Perhaps you're skeptical that two such disparate-seeming professions could share experiences? Let me give an example of what it feels like to solve a difficult puzzle after a long period of concentration:

At one moment I had none of this; at the next I had all of it. If there is any one thing I love . . . more than the rest, it's that sudden flash of insight when you see how everything connects. . . . I wrote a page or two of notes in a frenzy of excitement and spent the next two or three days turning my solution over in my mind, looking for flaws and holes . . . but that was mostly out of a sense of this-is-too-good-to-be-true unbelief. Too good or not, I knew it was true at the moment of revelation.

I've had that same feeling many times—it's the very best part of doing science. Except the text above is not from a scientist—it's from Stephen King when he figured out how to finish his book *The Stand*.[2]

So that was the first discovery—finding that doing science shares a surprisingly deep connection with other creative arts. Some background reading taught me that others before me had come to the same conclusion, finding parallels between doing science and writing novels or poetry, painting, composing music, and more.[3] That was not a disappointment, quite the opposite. Knowing that I wasn't the only scientist to have converged on this point spurred me to explore the connections further.

A little more sleuthing, this time on the science side, and it became clear that many other scientists experience the mysterious, inspirational, and often chaotic side of doing science. I'll let some other scientists speak for themselves—here's a sampling taken from specialists in immunology, biology, neuroscience, physiology, and psychology:

[S]cientists should not be ashamed to admit, as many of them apparently *are* ashamed to admit, that hypotheses appear in their minds along uncharted byways of thought;

that they are imaginative and inspirational in character; that they are indeed adventures of the mind.[4]

—PETER MEDAWAR

Intuition . . . is something subconscious, which, all of a sudden, comes out of the clear sky to you and is absolutely a necessity, more than logic.[5]

—RITA LEVI-MONTALCINI

We seem to forget . . . that some of the most important discoveries have been made without any plan of research . . . that there are researchers who do not work on a verbal plane, who cannot put into words what they are doing.[6]

—CURT RICHTER

All sorts of things can happen when you're open to new ideas and playing around with things.[7]

—STEPHANIE KWOLEK

We believe that such rules as to how science (with a capital S) is done, or should be done, are largely fiction, an attempt to retrospectively codify a process that often amounts to groping. There simply are no rules as to how to do science.[8]

—DAVID HUBEL AND TORSTEN WIESEL

By some unspoken rule, a scientist's feeling of awe for the natural world must be kept under wraps; to acknowledge wonder is tantamount to unreason, and therefore treason.[9]

—SARAH LEWIS

These sound like the musings of poets, not scientists. This is not to suggest that scientists don't eventually work within the confines of a fairly restricted set of rules—what you might call the grammar of science. But this rule-governed behavior may come fairly late in the process. It all starts with an idea about what you might study and how you might study it. To return to the writing comparison, or any other creative endeavor for that matter, as I see it there are two main challenges to success. One is mastering technique. The other is coming up with good ideas about where and how to apply that technique. You may stock a lab with scientific instruments, but there is no ideas store.

Tom McLeish puts it concisely in his book on scientific creativity,[10] suggesting the process of science has two stages—first the conception of an idea, and second the testing of that idea. McLeish is quick to point out: "Look up any popular definition of 'scientific method'—it is exclusively to this second stage that it refers." I'll add that most books on the topic of doing science also deal exclusively with the second stage, leaving out the source of ideas (more on this later). Here again, a comparison with writers is apt. My favorite quote on the topic comes from Stephen King in his memoir on writing: "We are writers, and we never ask one another where we get our ideas; we know we don't know."[11]

That said, whether you are a writer, painter, musician, photographer, or other creative artist, there are certainly some best practices that set the stage for the emergence of new ideas. Stephen King has much to say on this topic when it comes to writing (*his* muse happens to live in the basement, which seems appropriate).

The same is true for science. One of my main goals in this book is to describe some of the ways to set the stage for new ideas and discoveries. In my own case, the intellectual leap from subjective experiences and muses to the concrete world of best practices required turning from Stephen King and his fellow writers to the science philosopher Thomas Kuhn and his landmark work on scientific revolutions. I was influenced not by Kuhn's famous account of revolutionary science and paradigm shifts, but rather by his insightful description of normal day-to-day science, which resonated with my own experience in the lab. That, along with Kuhn's interpretation of what motivates scientists and how experiments in a particular field change over time, helped reveal an underlying pattern in the seeming chaos of many of my own investigations. The pattern (which I describe in chapter 1) is pervasive, at least for me—it underlies most of my discoveries.

That is the origin story for this short book—you might say my recipe included a few sprinkles of electric eels, tentacled snakes, zombie-making wasps, and star-nosed moles, then I added a dash of Thomas Kuhn and a pinch of Stephen King and stirred. Many unexpected insights emerged from the stew. This book also serves as companion to my previous book, *Great Adaptations*, giving the "how" for a series of discoveries. That said, I had to wonder how much of my process, exploring extreme adaptations at the fringes of biology and neuroscience, would be useful in the wider world of science.

Here again books about writing were encouraging, this time not for their content, but rather for their sheer number and diversity. If you do a little searching, you will find that aspiring writers have vast resources upon which to draw for advice and inspiration—my shelf holds over two dozen books on writing. When I asked a friend about their own collection, they sent me a photo of their bookshelf because the list was too long to type. These books are like candy to anyone contemplating a writing project.

What could possibly be useful about so many different perspectives on writing? It allows people to adopt a strategy that I believe is key to success in almost every walk of life. Namely, you can learn about the vast diversity of successful approaches to a difficult endeavor, reject those that don't align with your own skills and personality, and adopt the approaches that work best for you. For example, some writers insist that outlining a project ahead of time is essential (Jon Franklin); others reject this strategy as a constraint on the imagination (Stephen King). As you might imagine, the pool of ideas for how to approach a writing project is broad and deep.

That same selective process—that is, finding out which strategies and approaches work best for you—applies to science as well. The problem is, there are far fewer accounts of how science gets done. And by this I don't mean books about data analysis, framing hypotheses, grant writing, lab management, or the statistical basis of experimental design. Many such books exist, but they mostly focus on what Tom McLeish called the second stage of science—the testing stage. What about the more personal account of how scientists come up with ideas, approach new problems, and stay motivated—not to mention what goes wrong and how they fail? I have only a handful of such books on my shelf, and it's telling that one is from the 1800s. To mention a few of the few, there's E. O. Wilson's *Letters to a Young Scientist* (2013),[12] Medawar's *Advice to a Young Scientist* (1979),[13] and Santiago Ramón y Cajal's ancient, but still popular, *Advice for a Young Investigator* (first published in 1897).[14] There is clearly a need for more perspectives on how to approach science. Hence my decision to persevere and put some of my own practices down for the record.

I will often compare the process and experience of doing science with that of writing. Sometimes it's not a comparison per se, because scientists must write (see chapter 4). But even outside the writing domain, there are many analogous challenges and solutions shared by these two creative endeavors. I think the ideas are more tractable when you can see how they apply equally to such seemingly different vocations.

At the same time, I will try to convey some of the specific practices and strategies that have helped me to solve scientific puzzles, design experiments, make unexpected discoveries, put discoveries in context, search for beauty in the data, and deal with failures along the way. Although I have included many different examples of studies that support the ideas I will present, the most detailed accounts and the majority of the figures come necessarily from my own work. I hope it will be obvious from what follows that I am not advocating for a particular approach to something as complex and diverse as science. Rather, my goal is to add a little something to the ideas pool. I hope you find something to absorb.

P.S. The QR codes that accompany some of the book's figures link to movies related to a figure or topic. Scan the code with your phone (or, in the ebook, click on the code block) to access the videos.

1

Discoveries

YOUR ATTENTION IS VALUABLE. It's no secret that attention is a much-coveted resource, to the point that many companies use sophisticated computer algorithms to harvest and sell that resource at every opportunity. At the heart of it, this chapter is about the value of your attention in science. That said, there will be no admonitions about how to spend your time. Instead I hope to identify some of the things that draw and hold our attention when it comes to scientific problems. It is possible to solve scientific problems without paying close attention to the data, and you might even have a successful career in science without being fully engaged. But without paying close attention you risk missing one of the best parts of being a scientist—the unexpected discoveries that lie hidden in plain sight.

With this truism in mind, what follows is my attempt to describe an algorithm of sorts for making discoveries. I'll call it "the pattern." I'll introduce it in this chapter and build upon it in the next two chapters. You may find that what I have to say falls under the broad umbrella of common sense. But keep in mind that the entire scientific endeavor has been described as organized common sense,[1] and that "organized common sense" is how we discover incredible things. In the same vein, laying out the simple pattern for how discoveries have unfolded has allowed me to identify some unexpected facets of the scientific process. For example, as you will see in chapters 2 and 3, the pattern I will describe reveals that two seemingly unrelated parts of science—namely, experimental design and the search for art—can each have the same effect, sometimes even a combined effect, on the process of discovery. The pattern I will describe also helps to emphasize the pragmatic role of technology; after all, microscopes, telescopes, slow-motion cameras, and many other instruments are all means of focusing attention on a particular aspect of our world and universe.

But let's start by leaving lab coats, test tubes, and all the trappings of the laboratory behind, and instead move to a forest. This is, quite literally, a more natural setting for considering discoveries. It was my childhood training ground, and the same is true for many other biologists. Though you don't need to be a part-time naturalist to appreciate this analogy to scientific research—consider Nobel laureates David Hubel and Torsten Wiesel, and their approach to neuroscience: "David and I approached the visual cortex as explorers of a new world. . . . At times we felt more like naturalists of a bygone era."[2]

So you're at the forest's edge on a nice spring day, armed with binoculars. You want to discover some interesting wildlife. Where do you start? Despite the theoretical utility of the strategy (or its use by the Coast Guard to find survivors at sea), nobody stands at the forest's edge, divides the scene into a grid, and dutifully scrutinizes everything in view, one segment at a time. In fact, selective attention is so integral to our nature, a robotic search strategy wouldn't even occur to you. And anyway, you'd be hard pressed to carry through such a ludicrous search without being distracted by the very thing you had set out to find. A hawk may appear overhead, harried by smaller birds protecting their nest, or deer might cautiously emerge from the tree line, or (always the attention-getter) you might hear the distinctive sound of a snake slithering through nearby leaves. The point being, even if you started an undirected search, your attention would be selectively directed to the sights and sounds around you.

What, then, is the key to success for finding wildlife? It all starts with a single ingredient—curiosity. Everything else will flow from there. You will begin to explore your surroundings, perhaps taking a path into the woods. As you move through the environment and encounter animals, learning will come naturally, without conscious effort. Over repeated visits you may eventually discover, as I did, that life concentrates in transition zones, where forest meets field or rocky slope. You may also learn that water is a universal attractor, as is sunshine— especially in the spring. Look on the rocky slope, in the sun, between the forest and the stream, and discoveries abound. A minute's walk uphill onto the dry, shaded forest floor, and there's orders of magnitude less to find. If you explore in this way, you will start to have questions. Let me give an example that brings the analogy closer to science.

When my brother and I were children, we were both obsessed with turtles. As we explored the forests and streams around our house, we checked off one species after another from our field guide, the way bird-watchers might compile sightings in their life list. But try as we might, we never could find a wood turtle. Still, we never forgot our quest, and years later we resumed the search (it helped to have a driver's license and a car). We read books and papers about wood turtles, and learned that wood turtles hibernate by sitting at the bottom of streams and rivers during the winter. Next we consulted topographic maps and, having found a promising stream in western Maryland, we packed some food, coats and gloves, and our wading boots and headed out on a mission.

When we got to the stream, we discovered that many of the slower, deeper parts were covered by ice. Undaunted, we made our way carefully through the shallow, faster-flowing parts, avoiding the ice and deeper water by occasionally scrambling along steep banks. Somewhere along the way, we each took a minor fall that filled our knee-high wading boots with freezing water. But we kept at it, partly drawn along by the thrill of false alarms—what we called "rock-turtles," which (like stick-snakes) look remarkably lifelike until you get close. And then, to our mutual delight, we spied the real thing—a hibernating wood turtle, covered with silt, sitting motionless in the cold water among the scattered rocks. After seeing the first one, we had a better search image and soon saw a half dozen more. Our quest had succeeded in a single day. But that's not the important part of the story.

Partway through our precarious hike, my attention was drawn to an unexpected movement. I watched, in utter astonishment, as a wood turtle slipped from the sunny bank into the water and crawled to safety under the ice in the deep part of the stream. Mind you, I grew up confident in the acquired knowledge that turtles simply do not sun themselves, or move about for that matter, when the temperature is in the 30s. Here was a personal discovery that drastically revised my view of the limits of turtle metabolism and behavior.

None of what I have told you is groundbreaking in the world of professional herpetology.[3] But I hope this vignette of personal discovery gives a taste of how some discoveries unfold. You may start from ignorance, but if you're curious, you cannot help but learn by interacting with your study-system, in this case wildlife in a forest. As you become an expert, questions and puzzles emerge and are pursued with focus and passion—hence our winter hike in the middle of nowhere with freezing, sodden feet. Without that singular focus of attention, driven by the unsolved puzzle of where to find a wood turtle, we'd never have seen a sunbathing turtle on the edge of an icy stream.

That pattern—moving from ignorance to expertise; from expertise to a specific, motivating puzzle; and then from that puzzle to something unexpected and extraordinary—is the same pattern that has played out many times in the course of my laboratory work. Next I want to consider how this pattern compares to the historical accounts from other fields of science, especially as suggested by (for example) Thomas Kuhn's account of progress in the physical sciences.

Finding Structure

Thomas Kuhn (1922–96) was one of the most influential philosophers of science in the twentieth century. For those not familiar with Kuhn, he was trained as a physicist at Harvard in the 1940s before switching to study, and teach about, the history of science. In 1962 he published his landmark work *The Structure of Scientific Revolutions*.[4] You might say the book produced its own revolution in the way scientific progress is viewed. Among his most famous concepts is that of scientific "paradigms" that are periodically replaced during scientific revolutions (later to become the ubiquitous, and overused, concept of "paradigm shifts").

I'm most struck not by Kuhn's famous concept of scientific revolutions, but rather by a more subtle part of his account— what might be aptly called a typical scientist's day job. Kuhn made a compelling case that most scientists, most of the time, are not involved in the kind of extraordinary science that leads to revolutions of theory or paradigm (e.g., Einstein's theory of relativity supplanting Newtonian mechanics). Rather, their typical work is to engage in the less appreciated, inglorious business of solving small puzzles within their restricted area of expertise. This seeming demotion from revolutionary thinkers to puzzle-solvers was apparently insulting to many scientists.[5]

Not so for me—speaking for one scientist, an insider if you will, Kuhn's bird's-eye view of the process seems remarkably accurate. I can only embrace the moniker "puzzle-solver," given that, as you will see, I have invoked the same metaphor to describe my own behavior and motivations as a scientist on several occasions.[6] On the face of it, this description of scientific behavior might seem an inconsequential matter of word choice. But words matter. In this case the words help to highlight common behaviors and, equally important, common motivations. Everyone has felt the rewards of solving a puzzle, and an unsolved puzzle can become an obsession. Let me offer an example you may find amusing.

The Power of Puzzles

Not long ago, I allowed an electric eel to leap up and shock my arm as part of an experiment.[7] When the paper describing the results was published, it raised some eyebrows—why would I do such a thing? Scientific papers don't provide the thought process behind most experiments, but I got the chance to explain myself recently—here's what I said in my book *Great Adaptations*:

> Have you ever worked on a puzzle only to find at the end, rather than the satisfaction of the complete picture, there is instead a glaring hole in the scene? You'll move furniture and look into floor vents until you find the missing piece. In our case, it's inevitably the cat's fault. We do one puzzle a year, aiming to finish on January 1 (a recent habit inspired by my parents). It usually takes a few days of casual effort if the cats behave. Imagine what it would be like to work on a puzzle for an entire year, only to find a missing piece in the very center of the most interesting part. That's how I felt after working on the circuit that forms when an electric eel leaps up to attack. . . . There was an obvious solution. . . . I would need to use my own arm.

It's telling that I converged with Kuhn and used the puzzle analogy to express my motivation as a scientist. The problem at hand was an attempt to characterize the circuit that develops when an electric eel attacks a terrestrial animal by leaping out of the water. The solution required the assembly of comically complex and esoteric equipment—as expected for such a narrowly focused question. The last variable could only be solved with certainty by giving the eel a realistic biological target—hence the use of my arm (see figure 1.1).

The experiment was a success, though there were no surprises—the recorded electrical current through my arm fell within the predicted range.[8] But that wasn't disappointing—as with a complicated jigsaw puzzle, knowing what the answer will look like ahead of time does not detract from the satisfaction of finally completing the picture.[9] (And it's not just me; the reviewers and editors at the scientific journal where I submitted the results were also satisfied.)[10]

I offer the eel experiment as an entertaining example of a puzzle-solving scientist doing normal science. In Kuhn's words (paraphrased for gender neutrality):

> Though the outcome can be anticipated, often in detail so great that what remains to be known is itself uninteresting, the way to achieve that outcome remains very much in

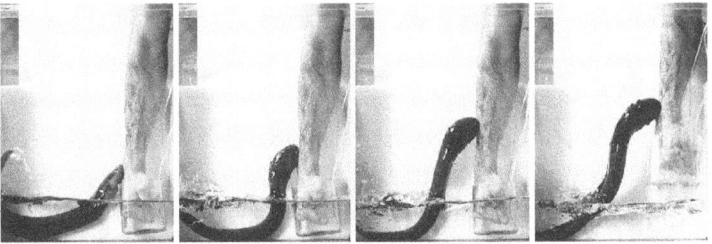

FIGURE 1.1. Images from an experiment measuring the amount of electric current that flows through a human arm (the author's) when an electric eel emerges from the water to attack.

doubt. Bringing a normal research problem to conclusion is achieving the anticipated in a new way, and it requires the solution of all sorts of complex instrumental, conceptual, and mathematical puzzles. The person who succeeds proves themselves an expert puzzle-solver, and the challenge of the puzzle is an important part of what usually drives them on.[11]

Notice that in the eel example above, there was no unexpected discovery. In other words, despite my obsession with the puzzle and my focused attention, I did not find anything special hidden in plain sight. That is a common outcome for an experiment. For now, I am emphasizing the first part of a pattern of discovery—the attention-focusing properties of a scientific puzzle.

It's hard to overstate the lengths to which scientists will go to solve their puzzles. Consider Marie Curie's single-minded work to isolate the element radium (quoted from her biography[12]): "Marie continued to treat, kilogram by kilogram, the tons of pitchblende residue which were sent to her on several occasions from St. Joachimsthal. With her terrible patience, she was able to be, every day for four years, a physicist, a chemist, a specialized worker, an engineer, and a laboring man all at once." I cannot help but add that she did this work either outside or in an uninsulated shed, exposed to freezing temperatures in winter and sweltering heat in summer, all the while protecting her equipment from leaks in the ceiling whenever it rained.

For a more recent example, consider scientist Katalin Karikó and her decadeslong obsession with messenger RNA (mRNA). After many years of daily work, facing constant resistance from the mainstream scientific community, her efforts finally paid off. She and her colleague Drew Weissman eventually found a way to harness mRNA, forcing the body to make novel proteins, and so laid the groundwork for the COVID-19 vaccines that have saved countless lives.[13] (I can think of no one more deserving of a Nobel Prize.)

The main point in highlighting this behavior is not the sheer physical work that is sometimes required for a solution. Rather, it is the prolonged concentration of attention, on what is often a tiny sliver of nature. In fact, there is a natural progression in science such that, for any given field, the range of puzzles narrows over time, and questions become more focused and esoteric.[14]

This, in turn, leads to the second main ingredient for solving many puzzles in science—more elaborate and task-specific experiments. In other words, solving puzzles in science usually requires focused attention and, as a corollary, the design of uniquely focused experiments.

My own experiment, described above, typifies this circumstance. How *does* one measure the current that flows from the head of an attacking electric eel, through a person, through the water, and back to the eel's tail? I can say with confidence that the apparatus I designed is entirely unique in the history of science, and moreover, the components for its construction were not even available to generations of scientists that came before me. You might say that the esoteric question I posed could only be solved with equally esoteric equipment.

I need only point to the Large Hadron Collider for a quintessential example of uniquely focused experiments in physics. The James Webb Space Telescope provides a similar example. And speaking of telescopes, they make for a particularly useful metaphor, or general "stand-in," for the design of new experiments.

In the case of a telescope, focusing attention and narrowing a view are literal, and have the positive connotations intended for the progression of many (though not all) fields of science. Consider that the first image shown from the James Webb Space Telescope was introduced to President Biden with these words: "Mr. President, if you held a grain of sand on the tip of your finger at arm's-length, that is the part of the Universe that you're seeing, just one little speck of the Universe."[15] The new telescope is a standout example of scientific progress, and it epitomizes focused attention on a narrow view.

In my previous book, *Great Adaptations*, I suggested that looking through a telescope is a good, general metaphor for conducting experiments.[16] With either a telescope or an experiment, things that are otherwise invisible are brought into view, and in either case the newly observed scene may contain the expected, the unexpected, or both. As with a telescope, experiments may be aimed at different phenomena, and technological advances are continually improving our view. In the next section I add this telescope metaphor to the puzzle-solving framework.

The Pattern

So far I've described scientific puzzles and associated experiments as the first part of a path to discoveries. Now I want to get more specific by extending that path with a fictional example. I've created an unconventional fractal for figure 1.2.

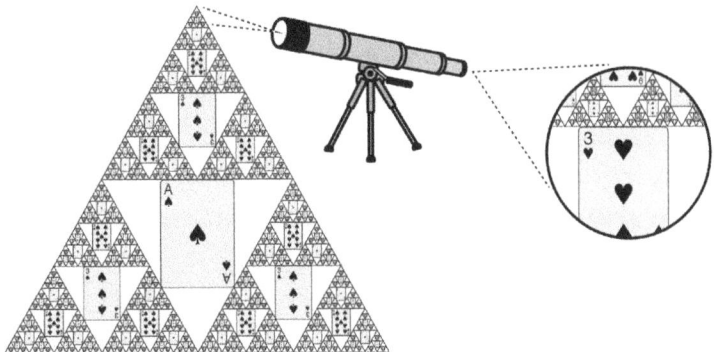

FIGURE 1.2. A fractal containing playing cards as an arbitrary puzzle. Does the numbering on the cards repeat at all scales, as would be expected? The telescope represents an experiment used to take a closer look at the system, providing the answer. The numbering repeats, but there is also an anomaly to discover.

Fractals are never-ending geometric shapes that repeat themselves at different scales, and they have always reminded me of the seemingly endless set of mysteries confronted by scientists with so many scales of the natural world to investigate.

In this case, the puzzle is: What happens to the numbering on the cards when you look more closely at a small area? Let's assume that modeling has predicted that what is seen at a glance in figure 1.2 will be true at every scale, meaning we will always see an ace, a three, and a nine. Does that actually happen? This question could stand in for many scientific puzzles—it could be part of a student's dissertation, a specific aim for a grant, the key to a Nobel Prize, or just curiosity. Whatever the reason, once an investigator has set her sights on the solution, the design of the required equipment for taking a closer look ensues.

As pictured at right in figure 1.2, experiments eventually bring the answer into view, and the model is confirmed, the puzzle of the numbers is solved—they simply repeat. Except, that's funny, the suit of the playing cards has changed. That discovery is not even part of the model. Worse yet (or better yet in science) the suit is anomalous—there aren't supposed to be black hearts on playing cards!

There's a reason people say the most exciting phrase in science isn't "Eureka!" but rather "That's funny. . . ." It is, after all, what Alexander Fleming uttered when he noticed a bacteria-free zone around some mold in one of his petri dishes.[17] Anomalies are usually interesting in their own right, but they may also suggest a whole host of additional experiments. (It took more than a decade from Fleming's observation of "bacterial inhibition" for others to finally purify penicillin for medical use.)

There is more to the example above than meets the eye. The sequence from attention-focusing puzzle to discovery is simple enough. But human behavior enters the picture when it comes to detecting anomalies—or perhaps I should say human perception. It's no accident that I used an unusual playing card to represent an anomaly. As pointed out by Kuhn,[18] playing cards were used in a set of now classic experiments by scientists Bruner and Postman in the late 1940s.[19] Their goal was to investigate the role of expectation when volunteers were asked to describe visual stimuli that were only briefly presented on a screen. Playing cards, being familiar to nearly everyone, were a natural choice for these tests, but the investigators added a twist. The cards included occasional "trick" or incongruous examples that had color and suit reversed—for example, a black three of hearts, or a red six of spades. As might be expected, subjects needed to see the trick cards for longer times (compared to normal cards) in order to correctly describe what they had seen.

The more interesting result was a strong tendency for subjects to report the anomalous cards as being perfectly normal—meaning they "saw" something different than what was presented. When, for example, briefly presented a red six of spades, they might report, with assurance, that the card was "the six of hearts." One subject reported a black three of hearts as being a "red" three of hearts for sixteen trials. Or, when presented with a black four of hearts, subjects might report—without hesitation—the card as being either a four of spades, or a red four of hearts.

It was only with longer presentations that subjects began to recognize something was wrong, often hesitating while seeming confused and distinctly uncomfortable. During this period, some might report the color of the suit as purple or gray. And as for shape, they might report they couldn't make out the suit. With increasing presentation time for anomalous cards, eventual recognition often occurred suddenly and with apparent shock. The study's authors referred to this as the "My God!" reaction—as, for example, when a subject said, "Good Lord, what have I been saying? That's a *red* six of spades." The lesson seems clear—expectation may hinder the recognition of anomalies. And what's true in the playing card experiment is true in science as well.

Before leaving off this section, I want to reemphasize how the metaphorical experiment above makes contact with the practice of science. As already suggested, the rewarding contingencies of puzzle-solving lead to focused attention and experiments that provide a magnified view of the subject. But in the real world, experimentation plays a second, essential role in revealing anomalies. Namely, in the practice of science, experiments are often repeated many times (see chapter 2). There can be preliminary experiments, then control experiments, then variations of experiments, then refinement and improvement of those variations, until finally enough data of sufficient quality have been collected for a confident conclusion worthy of publication. (The process is not unlike the revision of a piece of professional writing, which is rarely published in first-draft form.) To quote Nobel laureate Katalin Karikó, "While an individual experiment is the smallest possible unit of the research process, it is not itself research. In science, your overarching goal is to develop and test hypotheses; to do this you need results not from one single experiment but rather from a mountain of them."[20]

For the scientist, the inevitable result of this evolutionary process is repeated, progressively longer, exposure to the data. And these are *exactly* the conditions required for recognizing anomalies, as just described above. The latter conclusion is no minor point, as it has many implications for how scientists should approach data collection (more on that later). Next I'll illustrate this point with some personal examples, emphasizing attention-focusing puzzles and the value of clear and repeated exposure to the data.

Putting Some Cards on the Table

Perhaps you're skeptical of my description of doing science as similar to building telescopes to look at fractal triangles full of playing cards. How closely does this imagined scenario actually

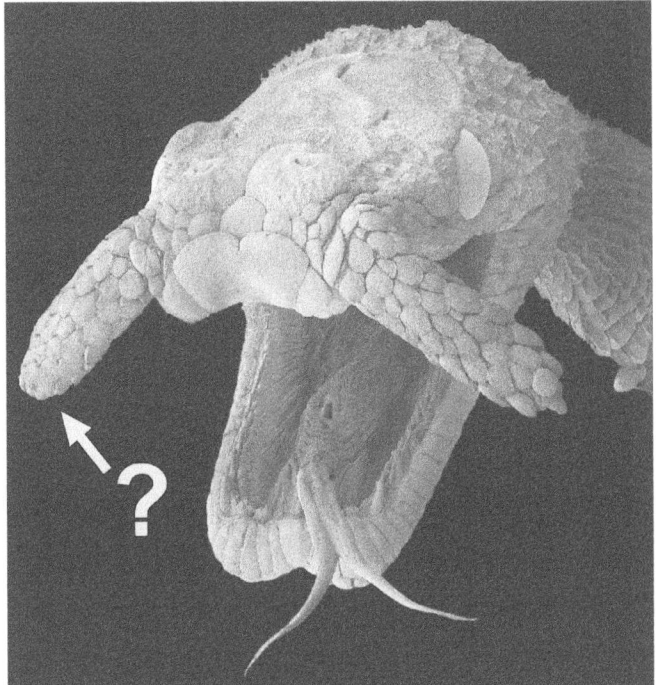

FIGURE 1.3. The unusual anatomy of the tentacled snake (*Erpeton tentaculatum*), showing the paired appendages.

parallel the behavior of a scientist at work? Answering for my-self, the pattern—from puzzle to experiment to discovery—is pervasive, at least when things are going well.

I'm going to give some examples of how this sequence has played out in my lab, starting with a simple example. Figure 1.3 depicts a scanning electron micrograph of the head of the ten-tacled snake (*Erpeton tentaculatum*). The puzzle is obvious—what the heck are those tentacles for? Herpetologists had wondered about this for over a century, but no one had figured it out. I decided to bring some of these enigmatic creatures to the lab to try to answer this question.

When considering this puzzle, it helps to know the snake is a fully aquatic piscivore, which is a fancy way of saying it eats fish. When hunting, the snake sits motionless in a J-shaped posture and waits for fish to come within striking range. Could the tentacles be lures used to attract fish, or perhaps water-motion sensors that detect fish movements? These two possibilities were high on the list.

To make a long story of many experiments short, we used a combination of behavioral, anatomical, and neurophysiological (nerve-recording) techniques to determine the tentacles are, in fact, water-movement sensors that help the snakes detect fish, especially in the dark or when the water is murky.[21] (The tentacles don't act as lures—fish rarely approach the tentacles, and when they do, the snake does not strike.) Determining the function of the tentacles was interesting, and it made for a nice publication, even though the answer was not surprising (again, many puzzles end with a satisfying solution even when the answer is expected). But the experiments that solved this puzzle required a long period of focused attention, especially on the snake's behavior, and that led to a discovery.

To tell you about it, I should first point out that tentacled snakes strike with exceptional speed, and fish are fast too—they can start their escape from a predator in less than 1/100th of a second (10 milliseconds). The whole life-or-death contest between snake and fish plays out in about 40 milliseconds, so watching the interaction took a very high-speed video system that records 2,000 frames per second. I was using

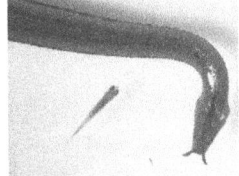

FIGURE 1.4. The remarkable "trick" used by tentacled snakes to take advantage of the fish escape response. The fish flees from the movement of the snake's body (middle image), and in the process swims straight into the snake's mouth (right side).

this system to watch tentacled snakes attack fish in slow motion when I noticed something astonishing.

Instead of swimming away from the snake's attack, most fish turned toward the snake's jaws and often swam straight into the snake's mouth (see figure 1.4). This was an anomaly to say the least—fish are renowned for their efficient escape response. (The escape response has been studied for decades and is the subject of over 100 scientific papers and several books.) If there's one thing fish excel at, it's detecting the sound of an attacking predator and swimming away. So what was going on with these snakes?

Further study revealed that tentacled snakes have what you might call an acoustic feint.[22] About 2 milliseconds (1/500th of a second) before the snake strikes, it moves a part of its body on the side of the fish opposite to the snake's head. This causes a sound in the water, which the fish hears, triggering the fish escape response in the wrong direction (wrong from the fish perspective at least). The neural network that controls fish escape commits to the wrong turn, and when the snake

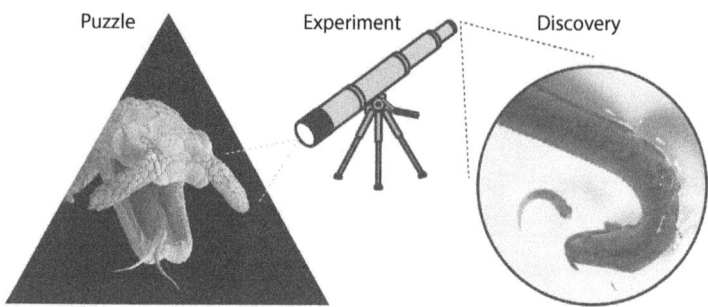

FIGURE 1.5. Schematic of the path to discovery for the tentacled snake's deceptive strike. Exploring the puzzle of the tentacles led to the unexpected attack strategy.

follows up with the strike, the fish cannot turn back. Instead, it swims straight into the snake's jaws. It's one of the most devious tricks in the animal kingdom, because the snake takes advantage of a neural circuit that's supposed to save fish from predators.

Figure 1.5 shows the events that led to the discovery of the tentacled snake's remarkable hunting strategy. It's a simple pattern, but there are many lessons embedded in each stage of this sequence. Consider the starting point. I would never have taken a closer look at tentacled snakes were it not for their unusual appendages, and yet the tentacles themselves have nothing to do with the snake's deceptive strike. You might even say the tentacles were a means (a way of focusing attention) to an end (making a discovery).

Next consider the experiments represented schematically by the telescope in figure 1.5. Designing experiments takes much time, energy, and resources. (Even making videos in this case included arranging stroboscopic lighting, computers, a digital chart recorder attached to a hydrophone, and proper aquarium configurations for high-speed recordings.) If you recall the example at the beginning of this chapter, you could think of this stage as the winter hike along a frozen stream with boots full of cold water. That might seem like an extreme description of experiments, especially if you like to tinker, but no one spends weeks assembling esoteric equipment for no reason whatsoever, any more than you are likely to hike in a frigid stream in the middle of nowhere without a good reason. A good puzzle changes drudgery into high adventure.

There is also more to say about the discovery side of this sequence. You might think that "discovering" the snake's deceptive strike was a simple matter of watching a video. But this was not the case at all. Think back to the experiment with the playing cards and how hard it can be to recognize something that's *not* supposed to happen. Fish are not supposed to escape toward a predator. So it was hard for me to notice, and then believe, what I saw. Moreover, two other investigators before me had examined the strike of the tentacled snake, but they did not recognize the deception. Why? Not because I'm more skilled. Rather, their video equipment was older and slower, providing either normal-speed video (30 frames per second) or far fewer trials and lower resolution (at 500 frames per second) because old-style film had to be chemically developed in a darkroom. I would not have recognized the snake's strategy with the older equipment either. I was fortunate to be able to record many trials to a computer with a state-of-the-art digital camera at 2,000 frames per second. This, in turn, gave me longer and more frequent exposure to the data. These are exactly the conditions that allow you to recognize anomalous cards.

Notice that technology has entered the picture here. There is a close symbiotic relationship between technology and attention. You cannot, for example, pay any attention whatsoever to a galaxy billions of light-years away without a telescope, and you cannot see the details of a tentacled snake's strike without a slow-motion camera.

Finally, I should add two things about unexpected discoveries that emerge from puzzles. First, don't forget the original puzzle! The snake's tentacles act as fish-detecting sensors, and that was an important result.[23] And second, most unexpected discoveries reveal more puzzles, and this in turn often leads to more experiments and further discoveries. This was certainly the case for the tentacled snake's strike. For example, is the snake's deceptive strike learned from experience striking, or is it an innate, instinctive behavior? It turned out to be instinctive.[24]

Dealing Another Hand

Next, I'm going to reinforce the pattern I described above with another example. But I warn you, this one is creepy; it's a story of discovery from one of the most famous parasitoids on earth—the emerald jewel wasp (*Ampulex compressa*), which makes zombies out of cockroaches as food for its young. I brought the wasps to Vanderbilt University to star in my yearly "Halloween lecture." I'll add that I am not an entomologist, and so I entered this area of research in a state of ignorance. That said, I possessed the key ingredient for success—I could not have been more curious about this species, which has

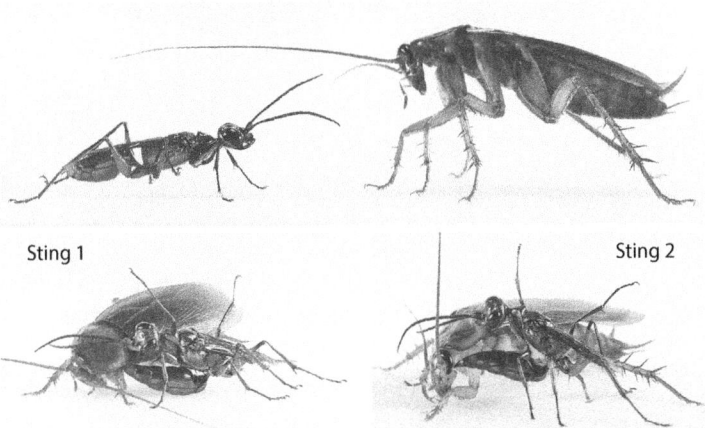

Sting 1   Sting 2

FIGURE 1.6. The emerald jewel wasp (*Ampulex compressa*) facing off with an American cockroach. Scientists have studied the two stings made by the wasp (the first between the front legs, and the second into the roach brain) since they were first reported by Francis Williams in 1942.

celebrity status in the genre of "creepy biology." In any case, I'll start by telling you about the wasp's spellbinding attack strategy.

The stinging behavior of the jewel wasp, and how it affects the cockroach nervous system, has been a focus of investigation for decades, resulting in dozens of scientific papers. Everyone who studies, writes about, or teaches about the jewel wasp (there's even a TED Talk that includes jewel wasps[25]) knows the wasp stings the cockroach twice, once between the front legs (sting 1) and once in the brain (sting 2) (see figure 1.6).

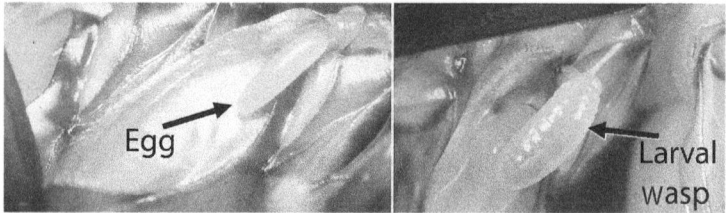

FIGURE 1.7. The typical location of the wasp's egg (left image) on the middle leg of a cockroach. The image on the right shows the hatched larva feeding on roach hemolymph after biting through a soft spot in the otherwise tough cuticle.

After that, the cockroach is never quite the same. Although the paralyzing effects of the first sting soon wear off, and the cockroach is able to walk, run, fly, or swim, if properly stimulated, it does none of these things. The second sting, into the brain, has zombified the cockroach. Instead of running for its life, the cockroach begins to groom itself, while the wasp goes off in search of a hole in which to entomb its victim. After about half an hour the wasp has usually found a suitable spot, and she returns to the cockroach.

She then grasps one of the antennae in her jaws, leads the roach to the hole, and pulls it inside. Next she lays a single egg on either the left or right middle leg and barricades the entrance with debris before departing. After a couple of days, the egg hatches and a tiny wasp grub punctures the soft membrane at the base of the leg and begins to suck hemolymph (insect blood) (see figure 1.7). After a few more days, the grub bites through the roach cuticle and crawls inside to feed on the roach's internal organs. It pupates within the roach's body cavity and about a month later an adult wasp emerges from the roach carcass, very much like the alien in Ridley Scott's classic science fiction movie.

A Puzzle

After the wasp has laid its egg, you might think the rest of the reproductive cycle is a fait accompli. But, in truth, the newly hatched wasp larva is in a precarious position—it can only survive if it finds the soft membrane at the joint of the roach's leg from which it can take its first meal. Finding that membrane, in turn, depends on the mother wasp's precision in laying the egg in just the right spot. Here's what I said about this conundrum in the closing pages of *Great Adaptations*:

> The mother's precision is a remarkable and underappreciated facet of the jewel wasp's biology. The egg is not simply left in the tomb, nor glued randomly onto the roach. Rather the wasp has evolved the instinctive ability to find just the right spot, on a nearly microscopic roach landscape, in the darkness of the entombing hole (though she actually has two choices, because the egg can be laid on either the left or right side of the roach). How does she do it? Although no one has yet determined which sensors are used, I've got my money on a group of hairs on the tip of her abdomen that look surprisingly like miniature whiskers.[26]

With that, I had found a new puzzle. The hairs stood out when viewed under the scanning electron microscope (see figure 1.8), and the obvious experiment was to cut them off and see what happened when the wasp laid her egg.

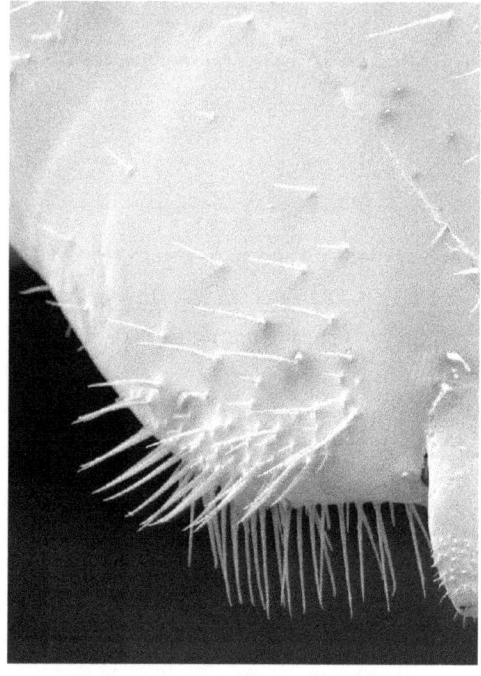

FIGURE 1.8. Tiny hairs at the tip of the wasp abdomen make contact with the roach as the wasp searches for the right place to lay its egg.

Focusing Attention

Solving this little puzzle is a lot harder than it sounds. For starters, just how *do* you shave microscopic hairs from a wasp's abdomen without getting stung or injuring the wasp? For this kind of operation, you must anesthetize the wasp, which can be done with carbon dioxide. With a little experimenting I found I could anesthetize an angry wasp for about a minute— long enough to at least try a microscopic haircut. Next, I ordered super-sharp scalpels made from volcanic glass. These

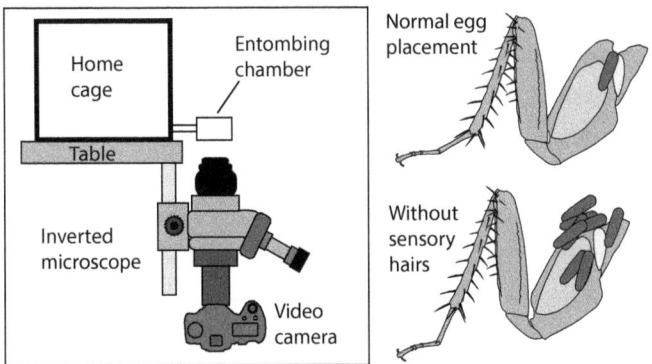

FIGURE 1.9. The left diagram shows the arrangement used to view and record the wasp's egg-laying process. The drawing on the right shows the effect of trimming off the sensory hairs.

are 100 times sharper than stainless steel scalpels. I practiced my shaving skills on dead wasps (using a microscope) until I could give a crew cut in sixty seconds, which was essential because I had to grip the wasp with my bare fingers—time was of the essence, to say the least.

And as for watching the wasp lay eggs, I solved this problem by flipping a microscope upside down and bolting it to a table. Then I made a series of clear-bottomed, Plexiglas chambers that served as artificial "entombing" holes for the roaches (see figure 1.9). With these positioned over the microscope, I was able to make high-resolution videos of the entire process— both before and after the hairs were trimmed.

The result? To cut to the chase, wasps were terrible at properly positioning the eggs without the sensory hairs on the tip of their abdomen. Normally, the egg is positioned so the larva hatches with its jaws right next to the soft spot where it needs to feed. But without the sensory hairs, the eggs were laid in all kinds of unusual positions, and later, when the larvae from these misplaced eggs hatched, they failed to find the "sweet spot" for their first meal. Instead, they starved to death while trying to chew through the roach's tough exoskeleton.

The result confirmed the importance of the mother wasp's behavior in properly positioning the egg and suggesting the hairs on the tip of the abdomen play a key role in the process. The results (which were not unexpected) made for a new, publishable contribution to our understanding of jewel wasp biology.[27]

The Black Diamond

Recall my suggestion that, in the practice of science, experiments are often repeated many times. This was epitomized in the study I just described. I had to closely watch every trial to keep the microscope in focus and properly aimed at the tiny scene. I had control trials and experimental trials, and I had to rewatch all the videos to make detailed measurements. Suffice to say I watched jewel wasps lay eggs more times than anyone else in history.

That's how I eventually noticed something strange. Before each female wasp laid her egg, she first extended the tip of her abdomen and probed the center of the cockroach in front of the middle legs. The wasp seemed to repeatedly fiddle around in this area, and in response, the cockroach usually extended its middle leg. This probing with the tip of the abdomen happened on every trial. Still, for the most part I ignored the behavior—it didn't seem important.

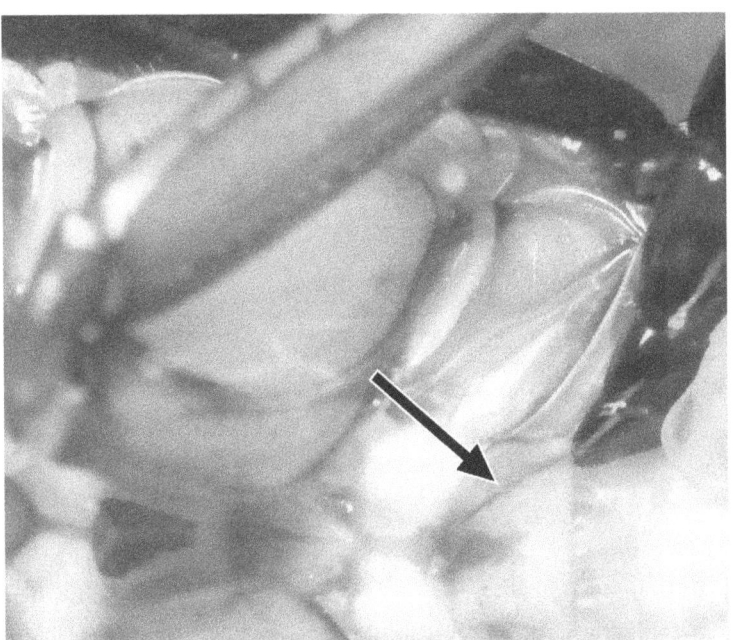

FIGURE 1.10. An image captured from video showing the newly discovered sting into the roach's central nervous system. The arrow shows the stinger under the cuticle.

Finally, though, I decided to take a closer look. That's when I had the "OMG" reaction described in the playing card experiment. The wasp wasn't just probing with its abdomen—it was stinging the cockroach. I could actually see the stinger extending below the roach's translucent cuticle—and it stung three times in a row (see figure 1.10).

To put this observation in context—I had just discovered that more than half of the stings made by this famous wasp had been overlooked since 1942, when Francis Williams published his first description.[28] Moreover, the function of these

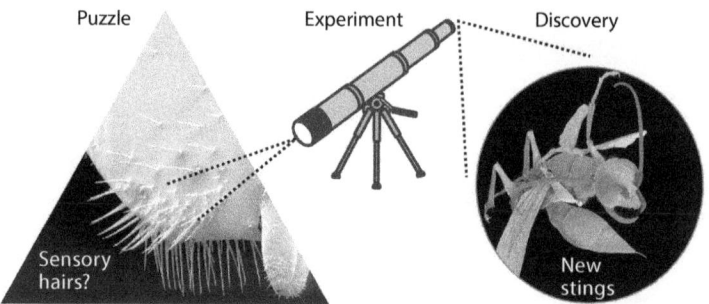

FIGURE 1.11. Schematic of the path to discovery for the newly discovered wasp stings. Exploring the puzzle of the abdominal sensory hairs led to the unexpected stings made in a new location in the cockroach's nervous system.

newly discovered stings, between the middle pair of legs, was both obvious and remarkable. They caused the middle leg to extend, making it easier for the wasp to lay her egg in the right spot. The finding reveals an added layer of "neural manipulation" by the wasp, and is a further testament to the astonishing sophistication of her attack—the female wasp targets three different areas of the cockroach nervous system in succession (first paralyzing the front legs with sting 1, next zombifying the cockroach with sting 2 in the brain, and then later forcing the cockroach to move its leg with three stings to the part of the nervous system that controls leg muscles).[29]

Discovering a new insect behavior obviously pales in comparison to discovering a new element or isolating a transformative antibiotic. And yet the pattern of this small science vignette could stand in for many others. Determining the role of the wasp's sensory hairs was the puzzle that focused my attention, leading to a small, incremental addition to our understanding. But that small puzzle led the way to a discovery that requires a substantial revision of the jewel wasp's well-known story (see figure 1.11).

Before leaving this case study, I want to reemphasize how aspects of my own behavior fit the pattern I described at the outset—identifying a puzzle, magnifying the view and focusing attention with experiments, and then making an unexpected discovery. I need hardly recount the telescope analogy when it comes to taking a closer look—the microscopic videography is a nearly literal example of this metaphor. But there is a key point to emphasize in this case.

It is hard to convey just how resistant I was to recognizing the new stings. The newly discovered stings were right in front of my eyes, as the saying goes, and recorded on video no less. Still, it took many views from several different angles before I could "see" what I was seeing. When I showed the videos to my wife, Liz (also a neuroscientist), neither of us could believe what we were seeing. I can't emphasize this point enough—both of us knew the established jewel wasp story too well, and our expectations were resistant to the upheaval. It was very much like seeing, but not recognizing, a black ace of diamonds—these are not supposed to exist. The lesson being, expectations can have a very powerful effect on our interpretations. As was the case with tentacled snakes, I was certainly not the first person to watch this stage of wasp behavior.[30] Here again, it was not a matter of skillful perception that led to the discovery. Rather, I happened to work on a puzzle that, by happenstance, focused my attention closely and repeatedly on the right place at the right time.

The Secret?

I could give many more examples of the pattern I just described, with different variations on the same theme. Having said that, I want to emphasize that discovering anomalies, or unexpected findings that fly in the face of expectation, is not something you can count on. It's the icing on the cake of science—an incidental side effect of conducting experiments and collecting data. And—importantly—it is not a requirement for success. In fact, one could argue that finding straightforward answers to a series of predictable puzzles—for example, completing an orderly series of specific aims from a grant—is the most efficient way to solve a circumscribed problem.

That said, for me, discovering something unexpected and pursuing it is the most enjoyable outcome of an experiment. The subjective experience is much like viewing an optical illusion during which your perspective shifts from one reality to another. Except, it is indescribably better when you are the one to unmask the illusion and see a new reality for the first time. That's why science, despite its dependence on logic and rigor, is full of mystery, plot twists, and feelings of awe.

Assuming I've convinced you that making discoveries is both exciting and useful, how do you do it? As you have seen from the examples I just described, paying close attention to the data for enough time is often the key, and this comes naturally if you are passionate about a particular scientific puzzle. But you can increase the odds of a discovery with other strategies as well. Opportunities abound with today's fast pace of technological innovations. To be more specific, if you are the first person to aim a new technology at a problem—even an old problem that you think has been solved—you may make an immediate discovery. Galileo simply aimed his telescope at Jupiter to discover its moons, forever changing our view of the universe. Scientists working with the James Webb telescope are in Galileo's shoes, and you can be too. It doesn't take a multibillion-dollar telescope—I had this experience with tentacled snakes by using a slow-motion camera—and (to emphasize the pace of technology) these days most people have a slow-motion camera in their phone. The inverse strategy can also work. Which is to say, you can aim older, well-established technology in a new direction.

Whatever your approach, I recommend staying close to your data. That, in turn, is how you get the best view through the metaphorical telescope called an experiment. But sticking close to the data is not as easy as it might sound. Since you already know the context, let me give an example from the wasp study I just described above. Much of the study was focused on the position of wasp eggs. It was tempting to save time and energy by simply measuring egg position post hoc, without observing wasps as they laid eggs. So why didn't I take this shortcut?

I could almost hear other scientists asking: "So . . . how do you know the egg wasn't shifted by movements of the cockroach sometime between when it was laid and when you looked?" (meaning that my measurements would not have been an accurate reflection of the wasp's egg-laying behavior). I would not have been able to answer that question if I had not observed and recorded the process. That's what I mean by staying close to the data. There are enough things in science that are impossible to measure—if you can be sure of something, why not be sure of it? Taking that extra step to be sure of the data pays many dividends, giving you confidence in your results and a better time with peer review, and if you're lucky, it may even lead to a discovery.

In the next couple of chapters, I'm going to describe some additional strategies that have kept me engaged, given me different views of scientific problems, and helped me to stay close to the data. But before leaving off this section, I want to give some more general, overarching advice about doing science. For this, I'm going to return to the comparison I made with writers in the introduction.

Professional writers are constantly asked about the secret to their success. My favorite anecdote on the subject is about author Natalie Goldberg. While she was giving a talk about writing, "someone asked her the best possible writing advice she had to offer, and she held up a yellow legal pad, pretended her fingers held a pen, and scribbled away."[31] In other words, the secret (write!) is so obvious it can be pantomimed. And yet, despite being obvious, the advice to "write a lot" is a mantra that needs to be repeated endlessly and is the bedrock of nearly every book on writing.

The same goes for science. It's harder to pantomime, but if there's a key ingredient to success in science, it's conducting experiments and collecting data. That's how you learn to recognize good puzzles, make progress on their solutions, learn from your mistakes, and make serendipitous discoveries along the way. And as with writing, this obvious advice is hard to put into practice. Yet that practice is, finally, where the best ideas come from. The best ideas come straight from the source— from the system under study. There's no muse in the wall of the lab, the muse is in the data.

2

Experiments

[E]xperimentation is a form of thinking as well
as a practical expression of thought.

—PETER MEDAWAR[1]

AT THE END of the last chapter I compared writers and
scientists—the two professions share much in common. Consider the evolution of a piece of writing. You probably know
that early drafts aren't your best work. You might think it's just
a matter of practice, but professional writers have the same
problem. To paraphrase Anne Lamott's advice—she's in favor
of "crappy" first drafts (I watered down her colorful descrip-
tor): "All good writers write them. This is how they end up
with good second drafts and terrific third drafts."[2] William
Zinsser puts it this way: "Rewriting is the essence of writing
well: it's where the game is won or lost."[3]

It is less obvious, but equally true, that many experiments
undergo a process of improvement and revision as they evolve
over time. And as with writing, the meaning, focus, and objec-
tive of an experiment may change drastically as a scientist tries
to describe a phenomenon or solve a specific problem. Many

writers find their story in the process of writing. Likewise, many scientists find their story in the process of conducting experiments.

And yet it seems this facet of the scientific process is under-represented in descriptions of how science is done. Formal accounts of the scientific method, especially those presented to students, tend to emphasize preemptive hypothesis construction, experiments designed to accommodate specific statistical tests, model building, theory, and the so-called hypothetical-deductive method. Published papers are notorious for omitting the often chaotic path that led to the final product. Readers are left with the impression that scientists usually come up with an orderly plan and then follow it to a natural conclusion. It is not unlike the idealized image of a professional writer producing flawless text in the first draft.

As a testament to this pervasive misconception that's close to home, I am asked to give a statistical justification for the number of animals I will study every time I write a university protocol (protocols are action plans for studying animals). The question presumes I know where my experiments will lead for years into the future. History has shown that this is rarely the case, and I doubt I'm the only one who feels uncomfortable when forced to make these imaginary projections. At such times, I am reminded of Richard Hamming's tongue-in-cheek advice: "In science, if you know what you're doing, you should not be doing it."[4] In other words, being overly confident about the outcome of experiments will divert your attention from the exciting anomalies in your data.

I'm not suggesting that theories, models, and statistics are unimportant. But in my own case, they enter the picture fairly late in the process. At the outset of a study, I like to keep it simple and travel lightly, minimizing the baggage of preconceptions. Simple experiments are precursors in an evolutionary process, either to die out if uninformative, or to be reproduced in varied and improved form if successful. And truth be told, most experiments do fail and die out (more on this in chapter 5). Stuart Firestein makes this point well in his book about science—aptly titled *Failure*. He suggests basic science is "just poking around most of the time."

Here again we come to the distinction between the two phases of science suggested by McLeish (and others before him[5]): first coming up with ideas, and later testing ideas. As with many other aspects of science, this distinction seems like common sense. Maybe so, but most published papers emphasize the testing phase of science, while the messy origin story may be entirely absent.

Writing on this subject, Nobel laureate Peter Medawar titled his provocative essay "Is the Scientific Paper a Fraud?" His suggestion was not that publications misrepresent facts, but rather, in his own words, "the scientific paper may be a fraud because it misrepresents the processes of thought that accompanied or gave rise to the work that is described in the paper."[6]

Getting access to the backstory is not easy, because scientists are human just like writers, and what writer would want the lousy first draft of their novel published alongside the polished final version? Even if they wanted to, I'd wager that most professional writers couldn't get their first draft published. Scientists have some of the same limitations and inclinations, often leaving a very skewed impression about the origin of ideas and experiments. It's a perfectly understandable state of affairs, akin to the problem of not publishing informative, but boring, negative results.

Understandable or not, this state of affairs can create some misperceptions. One is the impression that experienced scientists have the miraculous ability to come up with good questions by just sitting around and thinking hard. In truth, many of their questions arise through an interplay between experiments, results, and thinking. In the words of science philosopher Norwood Hanson: "Natural scientists do not start from hypotheses. They start from data."[7] Another, similar misperception is that scientists publish the first drafts of their experiments. In reality many experiments evolved from simpler beginnings, despite how the progression is portrayed in the final publication.

This being the case, beginners may start to ask themselves: "Why can't I just sit back and come up with good questions by thinking hard like everyone else?" Or they may try to emulate the most complex, final versions of published experiments too soon in the process, without considering that revisions, or even major changes in focus, may be required along the way. In the worst case, these two inclinations collide, and a poorly posed question is pursued with long and complex experiments, leading to epic failure. I speak from experience when I say failure is unavoidable and pervasive in science, so one goal is to fail well, which means with minimal loss of time and morale (the two are closely linked—see chapter 5).

I will reiterate my disclaimer, that other scientists may have a different, and perhaps more direct route to questions and experiments than I am suggesting here. That said, I'm going to describe how successful experiments typically go for me. I'll do that by illustrating the evolution of a set of experiments from my own lab because I can tell you when new ideas occurred to me. But before that, I want to recount a couple of case studies from other scientists, starting with the famous psychologist B. F. Skinner. He did us the great service of reconstructing his early behavior while a graduate student, and so we can see how his experiments evolved over time.[8] Plus, there's a much deeper and more profound lesson—Skinner's eventual focus on what maintains an animal's behavior also applies to what maintains a scientist's behavior.

Skinner's Case History in the Scientific Method

B. F. Skinner did his graduate work at Harvard, and as he put it: "So far as I could see, I began by simply looking for lawful processes in the behavior of the intact organism."[9] Although he officially entered the Department of Philosophy and Psychology, as with many graduate students, he spent the first year or so feeling out different lines of research and even different departments at Harvard. He was pulled in the direction of biology, but ultimately decided to stick with psychology. His reason was telling—in his autobiography Skinner says, "I was confirmed in my choice of psychology as a profession not so much by what I was learning as by the machine shop in Emerson Hall."[10] Soon he was spending most of his time at the shop, and one of the first things he made was what he called a "silent release box" for studying rats. This was essentially a

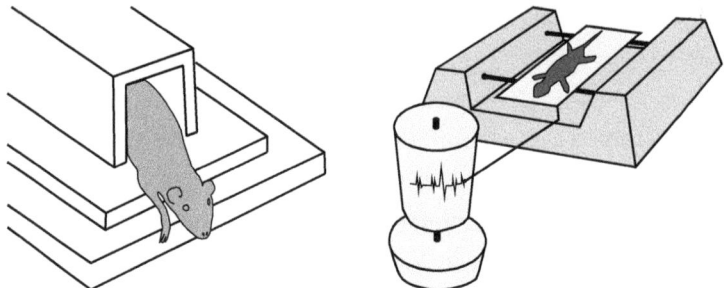

FIGURE 2.1. Two of Skinner's early designs for investigating the basics of animal behavior. The tunnel on the left was used to observe a rat's response to a soft clicking noise as it began to explore its environment. The apparatus on the right was used to record the movements of juvenile rats when their tail was stimulated.

chamber from which a rat could be released with minimal disturbance, after which it traversed a tunnel and came down a short set of steps (at left in figure 2.1).

His goal was to see how a rat adapted to a startling stimulus (a clicking sound) as it began to explore its environment. Rats are nocturnal and cautious by nature, so it took a while for the rat to peek out, look around, and eventually start down the steps. Then came the startling "click," and the rat's inevitable retreat back into the safety of the dark tunnel. Skinner waited patiently for the rat to reemerge before repeating the process, and eventually the rat stopped responding to the clicks.

This initial experiment was easy to design and kicked off his studies, but it wasn't very satisfying. He had to record his data by hand, and there were long delays as the rat huddled in the darkness. Eventually, his rats had babies, and he scrapped the first experiment and decided to study simpler behaviors in young rats.

The next apparatus had a small cloth-covered platform suspended by piano wire (at right in figure 2.1). On this he could place a juvenile rat, and then measure responses when its tail was pulled—the rats responded with a sudden leap forward. Now at least he had a consistent, repeatable behavior, and he added an automated data recorder (a so-called kymograph) that inscribed the movements of the platform onto a rotating piece of paper. The result was a straight line that showed each response over time.

The juvenile rat study was a step forward for recording data, but a step backward for anything that might apply to adult animals in everyday life.[11] It was time for another revision, so, building upon what he had learned, Skinner incorporated the best parts of the two previous experiments into a new apparatus for adult rats. This was an eight-foot-long runway that wiggled slightly when the rat moved (at left in figure 2.2). He gave the rat food at the end, but he also used clicks to occassionally startle the rat. The startle, which caused a corresponding vibration, was automatically recorded on the rotating paper.

Although data collection had improved, soon Skinner tired of carrying the rat back to the starting point after every trial. The obvious solution was to add a return path to the runway. He also moved the food reward to the corner of the return path (at right in figure 2.2). This made it easier for the rat, and also easier (and more rewarding) for Skinner, as he could sit back and collect data in comfort.

At this point Skinner noticed something unexpected that might seem like a minor distraction from his focus on the startle response. Namely, the rat paused for long periods,

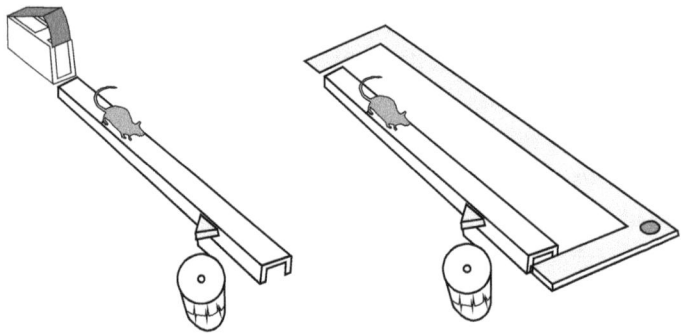

FIGURE 2.2. Two subsequent steps in the evolution of Skinner's experiments. Skinner retained the movement recorder (the so-called kymograph) but went back to studying the adult rat's startle response to sound, in this case on a runway (the runway was covered to form a long tunnel during experiments), as shown on the left. The drawing on the right shows the addition of a return path, with a food reward (later called a reinforcer), to reduce the experimenter's work carrying the rat back to the starting point between trials.

sometimes for up to ten minutes,[12] at the empty food dish. In Skinner's words: "There seemed to be no explanation for this." (Presumably, he said this because a rat would have gotten the most food by continuing along the runway back to the food dish, which would have been replenished.) He decided to abandon his focus on startle responses, and instead began to focus on the rat's response to food. It was a momentous change in the course of his research.

With his new focus on the rat's response to the food reward (he would later call food a reinforcer), Skinner began to further revise and automate his experiments. The next apparatus was an enclosed, rectangular runway that was elevated and balanced on a central bar. As the rat traveled around the circuit,

FIGURE 2.3. Further evolution of Skinner's experiments, now with a focus on the food (reinforcer). A complex combination of rectangular runway and automated food delivery was soon replaced with a single chamber with a lever. In the now famous Skinner box (at right), lever presses result in food delivery.

the entire runway tilted back and forth, moving a lever arm controlling a disk that automatically dropped a food pellet (at left in figure 2.3). At the same time, the data recorder was improved. So far, each of the rat's responses had been marked on a straight line as the drum holding the paper turned. But now the pen contacting the paper was automatically lowered each time the rat responded. The result was a curved data plot that was much better for showing rate of responding.

Finally, Skinner realized that a rat completing a circuit on a rectangular runway was an arbitrary behavior. After all, the runway might be shorter or longer—either would change the rate of reward delivery, but runway length was not an important variable—the rate of responding was key. So why not eliminate the runway completely? His solution was to make a box with a lever that could be pressed, activating a switch that operated the food delivery system. The result is the famous Skinner box (at right in figure 2.3). He had uncovered exactly what he had set out to find—lawful processes in animal behavior. What followed was an avalanche of new insights into the relationship between behavior and its consequences, or put more simply—learning.

This short history is in many ways the perfect example of the process I wanted to illustrate, because (ironically) Skinner was studying the consequences of animal behavior while experiencing the consequences of his own style of research (i.e., his own behavior). This was, of course, not lost on Skinner; in fact, it was his main point when recounting the evolution of his experiments.[13] Notice the relationship between his ability to efficiently revise experiments and the progressive improvement in his data as well as the increasing relevance of his results. It's a very rewarding experience, as I can attest. But just how rewarding is it?

Not long after he designed the Skinner box, the food delivery system jammed during one of his training experiments. It was a lucky accident, because the rat kept pressing the lever for some time before giving up. The resulting data plot (called an "extinction curve") showed a gradual diminution in frequency of lever presses. In other words, when the reward that followed the behavior was no longer delivered, the behavior slowly disappeared. It's a particularly lawful relationship between a behavior and its consequences—and probably something you have experienced.

Skinner described the data as "terribly exciting"—only a few years after entering graduate school his results were better than those of Pavlov (who had won the Nobel Prize). He goes on to say, "All that weekend I crossed the streets with particular care and avoided all unnecessary risks to protect my discovery from loss through my death."[14] That much excitement is a level above what you might call the Christmas morning effect—when you're so excited you can hardly wait for the next day. In other words, Skinner was enjoying his research and—to return to a more scientific parallel—there was little chance he would experience his own extinction curve in the lab.

Perhaps the most obvious lesson from this example (and the most comforting, for that matter) is that Skinner did not simply wake up one morning and sketch out an ingenious new plan, and experimental apparatus, for efficiently studying animal behavior (upper left in figure 2.4). Rather, it was a slow process of gradual evolution, during which Skinner took inspiration from his most recent results when designing each next apparatus (lower images in figure 2.4). He was certainly sticking close to his data, and his strategy epitomizes Medawar's quote at the beginning of this chapter—that "experimentation is a form of thinking."

There's another more practical lesson that may be less obvious. Skinner had chosen to work in a building with a nearby machine shop so that he could easily modify his own experiments. I said at the beginning of chapter 1 that designing experiments can feel like a winter hike with cold feet. Other things being equal, the easier the hike, the faster, and therefore farther, you will go. This is no minor point—an investigator's ability to modify an experiment is often the rate-limiting step in the course of its evolution. I'll come back to this point later, but first I want to tell you about the evolution of another set of experiments.

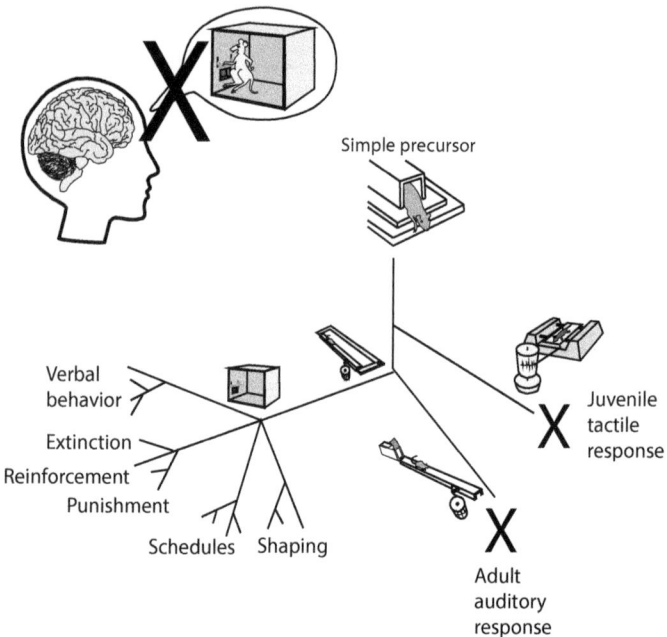

Simple precursor

Verbal
behavior

Extinction

Reinforcement

Punishment

Schedules   Shaping

Juvenile
tactile
response

Adult
auditory
response

FIGURE 2.4. The contrast between the common image of experimental design (top) and the reality for many experiments (bottom). Skinner's experiments evolved over time, shaped by the results of each previous iteration.

Turning Facial Recognition on Its Head

We humans are pretty smug about our large brains and all the amazing (and disturbing) things we have done with them to dominate the planet. Our status is in part the result of language and tool use, but it also depends on social bonds that allow us to work together toward a common cause. Part of that ability comes, in turn, from a very particular human expertise—a knack for quickly recognizing other human faces.

Learning to efficiently recognize individuals based on their face, and hence distinguish friend from stranger, or foe, is thought to have played an important role in human evolution—it's a part of the glue that holds us together. Moreover, brain mapping studies have revealed a specific area of the human brain, called the fusiform face area, that is essential for face recognition.[15] This is a remarkable thing—a specific module in our brains is devoted to learning faces—in the same way that the better-known Broca's and Wernicke's areas are devoted to language. This discovery suggested to many scientists that face recognition is so challenging it must be restricted to mammals with relatively large brains that have room for many large subdivisions.[16] That is, until a single investigator observing insects soundly dispelled this misconception.

In the late 1990s Elizabeth Tibbetts was a new graduate student at Cornell University, and she was in search of discovery. But how do you discover something new when it seems like everything has already been studied? Her first impulse was to explore a species that had never been investigated before, reasoning that anything she learned would be new to science.[17] With this plan in mind, Tibbetts set off to Zimbabwe in search of exotic African wasps. Unfortunately, political turmoil at the time made the work unfeasible. She returned to the United States and instead began to study a more common and accessible wasp, the northern paper wasp (*Polistes fuscatus*). These make their nests under roofs and overhangs in much of eastern North America and Canada. Her goal (working with her graduate mentor Kern Reeves at Cornell University) was to document the complex social interactions that take place in a paper wasp colony in order to test theories about the evolution of cooperation.

Most people are familiar with cooperation in honeybee colonies, where a single reproductive queen is the clear matriarch of the hive. She lays all the eggs, and the workers do, well, all the work. It's an orderly and, for the most part, peaceful arrangement. Not so for the northern paper wasp. This species has a much more contentious and even violent society, at least when the wasps first get together, because (unlike honeybees) multiple females create the colony and each has the potential to reproduce. As a result, the wasps fight it out to see who will be dominant, ultimately forming a strict social hierarchy—after which things settle down and become more peaceful.

Tibbetts was studying this dynamic in northern paper wasp colonies using a tried-and-true trick handed down for generations of bee and wasp investigators—each individual wasp was marked with a colored paint dot for easy recognition. Next the colonies were videotaped so Tibbetts could watch and score the myriad social interactions as the wasps formed their pecking order. Except one day, she accidentally videotaped a colony with two unmarked wasps. Not wanting to throw out all that useful data, she wondered if she could somehow tell the two wasps apart. If so, she would be able to continue with the analysis. As she closely scrutinized the video, she suddenly realized that each wasp had different and distinctive stripes and spots on its face.[18] In other words, she *could* tell them apart, based on their faces—no need for the painted spot on the back. And with that epiphany came a key question: Could the wasps tell each other apart the same way?

She followed this lead by first taking a much closer look at wasps' faces. She found a great number of distinctive features that could be used to distinguish individual wasps.[19] The next step seemed both obvious and simple: she could use the lab's stock of wasp-marking paint to alter a wasp's face and see how this affected social interactions. The result was clear: wasps with altered faces were treated as different individuals: and the fighting broke out anew, until the other wasps had relearned the face of their companion and social order was reestablished. (Control wasps were painted in areas that did not change their appearance, and these did not get roughed up.)

The finding was a sea change for our understanding of facial recognition. After all, the *entire wasp* is much smaller than the human brain area (the fusiform face area) used to recognize faces, not to mention the wasp's brain is 10,000 times smaller than ours. And yet the wasp is quite capable of this seemingly complex discrimination. It's one more example of how easy it is to underestimate animal abilities. And, as you might imagine, Tibbetts's results led to many more puzzles, questions, and discoveries. It turns out paper wasps have a "face filter" of sorts, such that they are especially good at remembering entire wasp faces, but are not so good at remembering equally complex patterns that don't resemble a (wasp) face—for example, when the antennae are missing.[20] The same is true for us: we key into certain patterns and find faces in clouds, slices of toast, rocks, and the like, when none actually exist. But we are not so good at recognizing an upside-down face, even though all the same patterns are there.

To return from the specifics of wasp faces to the evolution of experiments, a comforting lesson is repeated here (see

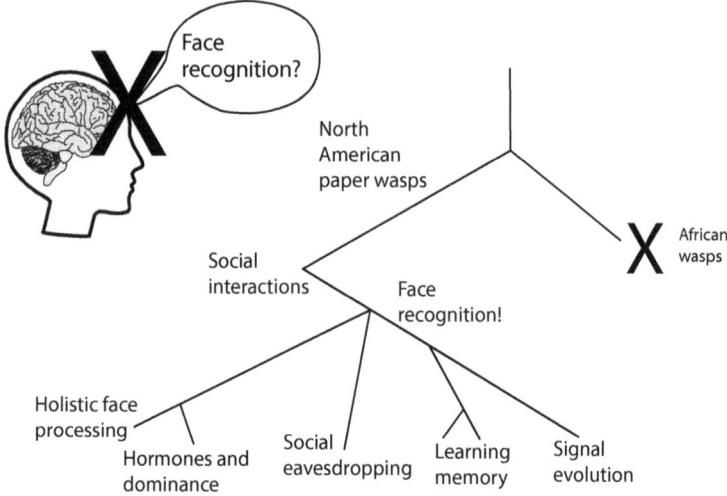

FIGURE 2.5. Another example of the contrast between the common image of experimental design (top left) and the reality for many experiments (bottom). The idea that wasps might be recognizing faces came to Tibbetts when she looked closely at the wasps in the course of experiments. Once this ability was confirmed, a new vista of studies and questions opened up.

figure 2.5). Tibbetts did not simply sit back, think hard, and come up with the exciting idea that wasps might be capable of facial recognition. Rather, the idea came to her in the course of experiments when, by happenstance, she needed to take an especially close look at her data. To put it more formally, she did not start with the hypothesis that *Polistes fuscatus* wasps could recognize faces; rather, she started with data. In the time since her discovery, Tibbetts has given this advice to graduate students: "The best scientists are prepared to take advantage of accidents . . . slow down and think about what you're doing, even when you're doing something that seems mundane."

Next I'll give an example of the evolution of one of my own experiments on electric eels.

Uncommon Sense

If a lab rat is the epitome of normal science, then perhaps an electric eel is the ultimate outlier. You might think of the eel as a living, breathing, swimming anomaly. (The electric eel—*Electrophorus electricus*—is an air-breathing freshwater fish, shaped like an eel, but unrelated to "true eels.") Everyone probably knows, or could guess, that electric eels use their electricity as a weapon. As if that's not enough, in 2015 I discovered that electric eels also use their high-voltage electrical pulses to track fast-moving prey—the way a radar-guided missile might home in on its target.

A key experiment from the published paper is illustrated in figure 2.6.[21] The figure shows an electric eel in its aquarium, after it has approached a large spinning disk that contains sixteen smaller circular disks embedded along the perimeter. One of those smaller disks is not like the others. Although they all look about the same (as a control for vision), one disk is made of conductive carbon (marked by the arrow) and the others are made of nonconductive plastic. The images are taken from a video that was made as the eel first detected the carbon disk using its high-voltage output, then chased it, and finally violently attacked the conductive disk, biting at it while giving off a volley of high-voltage electric pulses. (The QR code links to the full movie of the eel in action during this experiment.)

Why would an eel chase and attack a small conductive disk? Because the eel's prey (usually fish) are conductive. In other

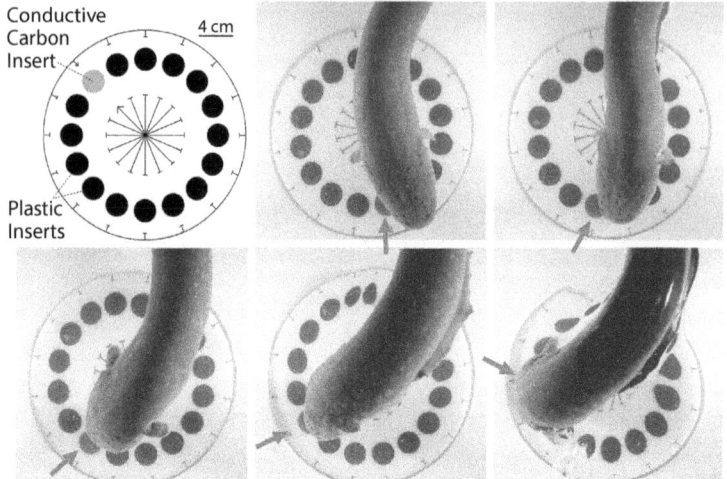

Conductive Carbon Insert

4 cm

Plastic Inserts

FIGURE 2.6. An electric eel attacks a fast-moving conductor (a carbon disk) hidden among similar-looking nonconductors, all embedded on a rotating platform. The experiment firmly demonstrated that electric eels use their high-voltage electric output to track prey. Prey are conductors, and the carbon disk thus has the electrical signature of prey.

words, the conductive carbon disk has the "electrical signature" of food. Think of this experiment as akin to using a laser pointer to play with a cat. The spot from a laser pointer may not look like typical prey, but it's close enough—cats love to chase small visual stimuli. It's the same for electric eels when it comes to small conductors. Cats are "visual" predators, whereas eels are "electrosensory" predators.

In fact, the experiment above showcases "active electroreception," or the ability of some fish to use electricity to probe their environment, thus detecting surrounding objects or other animals. It's a sense used by many fish that emit weak electrical signals, but the experiment above was the first to show that the stronger electrical output of an electric eel is, in addition to being a weapon, also part of a sensory system. I find the result astonishing, plus I think the spinning circle has a certain artistic appeal, as experiments go.

Artistic or not, I can assure you that I did not wake up one morning wondering whether electric eels use their high voltage to detect prey, and then set to work building the spinning circle test. But you would never know this from my paper on the topic. When I wrote up the results of this experiment for publication, I succumbed to the same compulsion as everyone else, leaving out the source of the ideas for the study. Here's all I said about my main question: "the possibility that the eel's high-voltage discharge also plays a sensory role has been overlooked. The experiments described below address this question."

It reads as if I sat around thinking harder than anyone else about electric eels and realized something had been missed by the scores of other investigators who had come before me (including Michael Faraday). But that's not what happened. So where did I get the idea? As usual, it all started very simply.

Simple Beginnings

I brought electric eels to Vanderbilt University for the class I teach in neurobiology and behavior (as with the jewel wasps). My goal was to get some good color photography and movies of electric eel behavior. The more I watched and photographed these amazing animals, the further I was drawn in, until I

FIGURE 2.7. Early steps in the evolution of experiments exploring active electroreception in electric eels. It all started with photography and then moved to recording the eel's electrical pulses while making slow-motion videos.

recognized some puzzles that I couldn't resist exploring. I've outlined the simple beginning in figure 2.7.

One of the most important functions of the electric eel's high voltage is to immobilize prey by causing sustained, involuntary muscle contractions, just like a law-enforcement TASER. But the eels also have a second, even more impressive trick. Namely, they can use a short blip of high voltage to cause hidden prey to twitch, which gives away their position to the eel. To study this behavior, I had to put a dead fish with working muscles into a plastic bag, thus insulating it from the eel's electricity. The fish was, in turn, attached by wires to an electrical stimulator that allowed me to generate a massive twitch. This gave me complete control over the fish muscles and allowed me to conduct a series of increasingly complex experiments exploring how electric eels respond to the water movements caused by fish. That's when I noticed something strange.

Usually, when an eel attacks a fish, it strikes while giving off a high-voltage volley and then bites at the fish. But for some reason, the eel never bit at the fish in the bag. Instead, when the fish twitched, the eel responded by giving off the usual high-voltage volley and the usual strike *toward* the bag. But it never

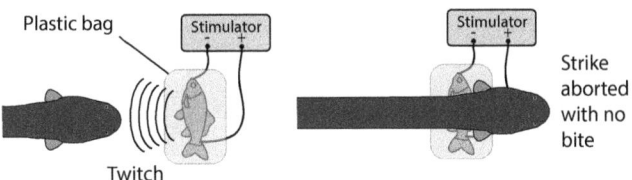

FIGURE 2.8. Electric eels strike out toward a twitching fish in an insulating plastic bag, but can't seem to locate the source of the water disturbance.

homed in on the exact position of the fish and tried to bite, as in a normal predatory strike (see figure 2.8). For some reason, the fish in the bag was "invisible" to the eel. How could that be?

That was the observation, the anomalous card if you will, that made me wonder whether eels might be using their high voltage simultaneously as a weapon *and* a prey-sensing radar system. This followed because the most obvious reason that a plastic bag would act as a cloaking device was if eels were *depending* on their high-voltage pulses to home in on the final location of prey.

With this idea in mind, as a "quick and dirty" test, I added a conductive carbon rod to the experiment (see figure 2.9), the idea being that a conductive carbon object has the electrical signature of an "uncloaked" fish. The result was dramatic— when the dead fish in the bag was made to twitch, the eel struck toward the fish initially, but then changed course and homed in on, bit, and tried to eat the carbon rod, as if the rod was a fish.

So now you have heard the origin story for the spinning circle test. The simple preliminary test with the carbon rod was a branching point, leading to the evolution of new

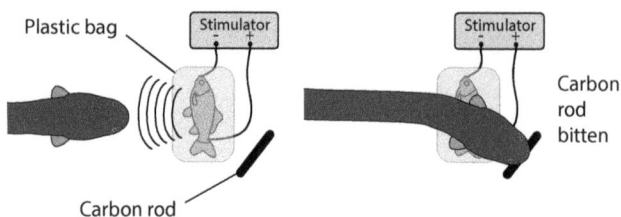

Plastic bag

Stimulator

Carbon rod

Stimulator

Carbon rod bitten

FIGURE 2.9. The first preliminary test for high-voltage electroreception. When a carbon rod (a conductor that imitates the electrical signature of a fish) was added to the previous experiments, the eel violently attacked, and tried to eat, the carbon rod—as if it were a fish.

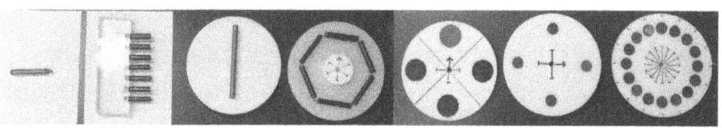

FIGURE 2.10. The evolution of a "branch" of experiments that led to the final and most elaborate test for high-voltage electroreception. The final stimulus configuration, with a single conductor hidden among fifteen nonconductors, is shown on the far right.

experiments. But there was no sudden leap from the carbon rod to the final spinning circle test. Rather, it was a gradual evolution as each experiment was revised and improved. This is something seldom documented, even in popular accounts of science, because, like early drafts of a novel, rough drafts of experiments are usually discarded. As it happens, I still have the evolutionary intermediates that led to the final "spinning circle" electroreception experiment (see figure 2.10).

Recall what I mentioned at the beginning: that simple experiments are precursors in an evolutionary process, either to

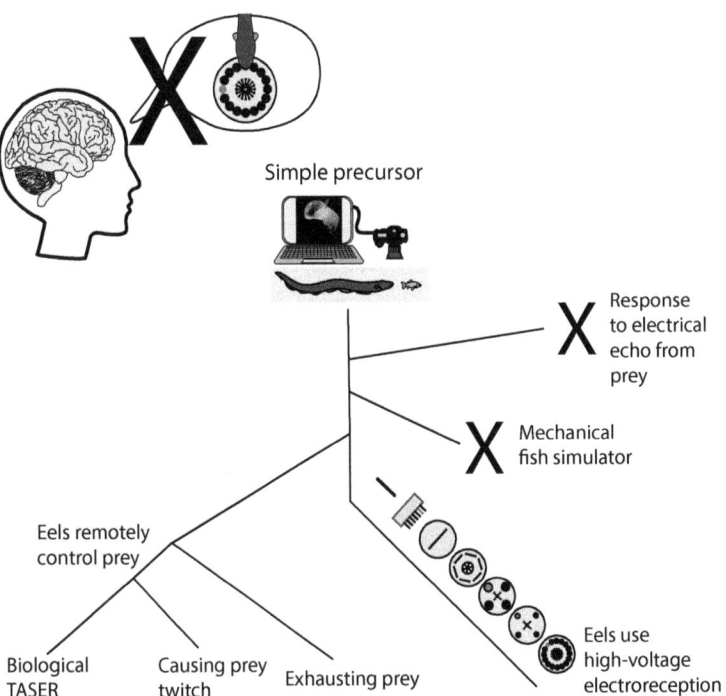

Simple precursor

Response to electrical echo from prey

Mechanical fish simulator

Eels remotely control prey

Eels use high-voltage electroreception

Biological TASER

Causing prey twitch

Exhausting prey

FIGURE 2.11. Another example of the contrast between the common view of experimental design (top) and reality (bottom), at least for the author. The simple precursor led to more elaborate experiments, including failures that were abandoned and successes that were pursued (not all successes and failures are shown). Each endpoint without an X resulted in a publication.

die out if uninformative, or to be reproduced in varied and improved form if successful. If there had been no response to the single carbon rod, the experiments would have died out. Instead, they survived and were refined as illustrated in figure 2.11, which shows their point of origin.

First Catch Your Trout

I did not simply think up the idea that eels might have high-voltage electroreception, despite how it sounds in the introduction to my publication.[22] Rather, the idea came to me while collecting data. So why didn't I explain the source of my ideas in my original paper? There is, unfortunately, little room or editorial interest in such "informal" details in formal scientific publications. Instead, the focus is on the question—the hypothesis if you will—and on the rigorous test of that hypothesis. Although there is an increasingly long list of required statements, disclaimers, forms, and declarations that authors must include when they submit a scientific paper, there is no requirement or even encouragement for authors to explain the origin of their ideas.

Perhaps the single most heavily emphasized requirement for a scientific publication is that experiments are repeatable. In other words, the recipe must be absolutely clear. Ideally authors will share not only the list of ingredients, but even the ingredients themselves (many journals require a commitment to share materials as a condition of publication). This is, of course, as it should be. But to continue with the analogy between cooking and science, I will quote the science philosopher Norwood Hanson, who used the same analogy, suggesting that many scientific accounts "begin with the hypothesis as given, as cooking recipes begin with the trout as given. In an occasional ripple of culinary humor, however, recipes sometimes start with 'first catch your trout.'" He goes on to point out that the typical account of experiments in physics "describes a recipe physicists often use after catching hypotheses."[23]

In my own case, as you have read, I most often catch my hypotheses while doing experiments. I find the interplay between the evolution of thoughts and the evolution of experiments to be key. And, this being the case, it raises a very practical dilemma. Namely, how do you learn the technical skills needed to design and modify experiments in the first place? This may require different strategies depending on your field, your personal experience, and your particular set of studies. I happen to be a bit old-school when working on a new experiment. I often use the most basic of tools—you might even say I got my start in the Middle Ages.

From the Middle Ages to Modern Technology

I grew up as part of what you might call the *Stranger Things* generation. When I wasn't out chasing critters in the local woods, I worked on my model train set, built model rockets, and, along with my brother and our assorted friends, I played Dungeons and Dragons in my parents' basement. The game was a gateway drug to the local Renaissance festival. Except visiting the festival wasn't enough; here was a chance to actually *be* a knight in shining armor. But how to get the job? There was only one thing for it: we would have to make our own weapons and suits of armor. So we started a miniature industrial revolution. By age fourteen, I had made a chain mail shirt by wrapping 16-gauge steel wire around a metal rod, cutting each loop off the spring-like result, and "knitting" them into a shirt using pliers. On my next visit to the festival, sporting my shiny new outfit, I was mistaken for one of the older actors and told to join the "mock" battle staged for the audience. It was my lucky break.

By high school we were all working at the festival for five dollars a day, though, as with Tom Sawyer's fence painters, we would have paid for the privilege. We were also improving our armor while learning about medieval history. As luck had it, my high school had an extraordinary metal shop. One room housed the standard workbenches, drill presses, table saws, grinders, and the myriad hand tools you would normally expect in any shop class from the '80s. But the second room contained a natural gas-powered forge, a foundry for melting bronze and sand-casting sculptures, oxyacetylene gas welders, arc welders, and more. This was heady stuff, and dangerous. Few high-school students had the drive to learn how to use such equipment, so the marvelously ill-conceived back room sat idle for years—until our crew showed up.

It was the perfect melding of obsessive motivation with opportunity. Our shop teacher, Mr. Brewer, had a stern façade that never quite melted, but I could tell he was secretly delighted to have us in the class. He finally had an excuse to use, teach about, and—most important—play with all that equipment in the back room. And so a delicate alliance formed and grew as we pushed the boundaries further and further. Eventually we had a production line turning out hand-forged spearheads, shield bosses, bronze axes, spiky maces, and even an entire suit of plate armor (it's still in my closet). It was the heyday of the shop, until it was all brought crashing down by the space shuttle.

In the early '80s the United States returned to manned space flight with the space shuttle *Columbia*. It harkened back to the Apollo moon program, when television sets were rolled into classrooms across the nation. In our case students and teachers crowded around a TV in the faculty office. During the countdown to launch, my friend Sean and I were working in the shop, he sanding and polishing a steel spearhead while I repaired a sword. We glanced at the clock and realized we were about to miss liftoff. With seconds to spare, instead of putting the weapons in a shop locker, Sean stuck the spearhead in his back pocket and I wedged the sword in my belt as we ran to join the crowd around the TV. We were just in time to see liftoff while standing directly across from the open door of the principal's office.

The rest of the day is straight from the script of any teen coming-of-age-story, with the added drama that Mr. Brewer was implicated. In short, after being ushered into the principal's office, we were told: "Having weapons in school and, worse yet, making weapons in shop class is absolutely forbidden. And though *you* might mean well, students have been known to stab or club each other with such implements." Our parents were called to escort us from the building. I'll never know what was said to Mr. Brewer, but he took it in stride, and probably stood up for us. After a brief hiatus, we were allowed back into the shop, but the arms race was over for good.

To return from medieval history to modern science, there are two lessons here; one is thematic and the other is purely practical. The thematic lesson is passion. Learning how to use tools in high school was a means to an end. And yet, because I was obsessed with my goal, I learned far more about using tools, and about medieval history for that matter, than I would have ever learned had I only been thinking about my grade in shop class. This general theme recapitulates the advice about puzzles, art, experiments, and anomalies—you will work much harder, learn more, and have more fun in the process, if you become enthralled with what you are doing. This emphasis on passion may be a cliché, but it happens to be true.

The second theme is far more practical. Namely, part of your toolkit in science might be, literally, a toolkit. In my own case, I cannot overstate the importance of being able to modify my experiments as the need arises—it has probably doubled the number of discoveries I have made over the years. And I'll add that, in science, a high ratio of discoveries to work is the very definition of fun.

Of course, these days many of the tools for modern studies have changed, and are likely to include such skills as writing and editing computer code, changing the conditions under which proteins crystalize, controlling a neuron with light, knocking out a gene with CRISPR, or any of a host of other techniques more typical of the modern laboratory (including the use of high-speed cameras). Moreover, much of what I do in my dusty backyard shop can now be accomplished on a pristine lab bench using a 3-D printer—with the added benefit that you don't need safety goggles and the odds of losing a finger are low.

I told my story not to recommend a shop class for students, postdocs, or assistant professors. Rather, I'm using my literal toolkit as a stand-in for many other forms of expertise that can be learned as the need arises. Here I'll return to Peter Medawar, who put it well: "The great incentive to learning a new skill or supporting discipline is an urgent need to use it. For this reason, very many scientists (I certainly among them) do not learn a new skill or master new disciplines until the pressure is upon them to do so."[24]

That said, there is a pitfall in many modern research environments. Namely, there is often such a myriad of support staff—including programmers, statisticians, microscopy technicians, histologists, geneticists, and machinists (not to mention experienced students and postdocs)—that it can be tempting to remain in the passenger seat, instead of learning how to drive. Most of these same professionals are willing to teach you how to do things for yourself. Don't remain passive, instead take every opportunity to get into the driver's seat, not only so you can optimize your future experiments, but also so that you can teach others when the time comes. I have a very long and growing list of driving instructors.

I hope I have convinced you that learning how to revise experiments is part of the thinking process. If that's not enough, there are other reasons. For example, when you outsource experimental design, you will almost certainly be shocked by the turnaround time. And there's another more important reason to learn how to improve experiments. Think back to Bruner and Postman's study with anomalous playing cards. It takes longer and more frequent exposure to the cards before people recognize anomalies. Experiments are your telescope for taking a longer and closer look at a question. The process of improving an experiment provides longer, different, and more frequent views of the data, giving you a much better chance of solving your puzzle while recognizing anomalies that lead in new directions.

The Polishing Stage

The advice in this chapter—that successful experiments often evolve over time—raises an interesting problem (a problem that's good to have). Namely, when do you stop revising a successful experiment and consider it good enough? I can only give you my own take on this question, and in doing so, I'll refer back to the challenge of writing.

Professional writers start with rough drafts and repeatedly revise. Eventually, they reach what is often referred to as the "polishing" stage—when the story is compelling but can be made more impactful and clearer. The same is often true for experiments—you might get the results published at an earlier stage, but it would be a better study if the experiments were given a final polish for clarity. That quest for clarity is why I kept improving the electroreception test illustrated with the spinning circle experiment. To return to the writing analogy, in Zinsser's words: "First clarity. If it isn't clear you may as well not write it, you may as well stay in bed. Clarity is the goal in writing, the main prize."[25]

It seems obvious the same holds for experiments—and when it comes to revising your experiments for clarity, there is far more at stake than appearances (see chapter 5). Improving your experiments is the best way to be sure you are onto something real. If, like a specious argument hidden behind muddled writing, your results evaporate as you view the effect with greater precision and from different angles, then be grateful you were saved from embarrassment. But if your result is real, improving your experiments for clarity (as with revising your writing) can pay many dividends—not least of which are an easier time with peer review and more attention after publication.

3

Art

The greatest scientists are always artists as well.

—ALBERT EINSTEIN[1]

WHEN I WAS IN GRADUATE school in San Diego, I worked on a project that required frequent trips across the country on Interstate 40. This took me past the Grand Canyon, and sometimes I'd visit the park along the way. On one of those visits, I joined a group walking along the South Rim, guided by a park ranger, and I'll never forget what he said. Over the years he had noticed that some visitors were overcome with awe at their first sight of the canyon and not sure what to do next. They had made long-planned road trips, or perhaps had come from overseas, but having finally arrived, the sheer scope of the vistas and incomprehensible distances could leave them at a loss. How do you interact with the Grand Canyon? Where do you even start?

His advice? Photography. For many people this was the answer. It gave them a way to focus their attention, literally. It was a reason to slow down and absorb the experience, instead of moving aimlessly from one overlook to another. It was a way to get drawn into the scenery—exploring details and beauty in the rock layers, in the clouds, in the tiny stream in the distance (which happens to be the mighty Colorado River). Even the courting ravens can be an unexpected source of artistic inspiration. For those who didn't already have a plan, photography gave the journey a sense of purpose and meaning. Plus it was a way to capture some of the grandeur and later share the journey with friends and family.

Scientific research can be a journey, too, a journey that's both awe-inspiring and daunting. And, ultimately, research is meant to be shared with your scientific family. With this truism in mind, I have a long-standing tradition of starting a new research project with art. Often this would be a photographic portrait of the animal I was studying, but I have also found compelling imagery in the anatomy of the brain, the growth cones of neurons, the development of embryos, the skin surface of a mole's nose, or even the standoff between a wasp and a cockroach. There's nothing like starting out of the gate with an inspiring image to boost your enthusiasm for the long, cold hike of experimental design. And there is more than aesthetics at stake—art is a secret weapon for making discoveries. But before turning to some of my own examples, I want to tell you about Spanish scientist Santiago Ramón y Cajal, who famously combined art and discovery to become the legendary father of modern neuroscience.

Butterflies of the Soul

Cajal (pronounced "ka-hal") was born in 1852, and to give some idea of the state of biology at the time, this was about a decade after it had been formally proposed that all plants, animals, and organ systems were made up of separate cells (so-called cell theory). In fact, Cajal's career would pivot around this particular question as it relates to the brain.[2] But there was no hint of Cajal's future in science when he was young; instead, by age eight he wanted more than anything to become an artist. He described his obsession with drawing as an "irresistible mania," and he spent every spare penny on pencil and paper. Sadly for Cajal, though luckily for the field of neuroscience, Cajal's father, a physician, had other ideas. He expected his son to become a physician, and he brought the matter to a head by having one of Cajal's drawings assessed by a local painter. The painter looked over the drawing for a suspenseful moment, and then pronounced a verdict—the boy has no talent.[3] History has shown this verdict to be a travesty. But it was taken as a final judgment by his father, who thereafter did everything in his power to discourage Cajal's artistic inclinations, instead setting him on a path toward medicine.

Cajal had many adventures and hardships along the way to becoming an established physician, but I will jump forward to the 1880s, when he decided to take the path of histology—the study of microscopic tissues. Advances in techniques for staining and cutting tissue had given increasing support to cell theory, but the fundamental units that made up human and animal brains remained an enigma. Were the neurons (cells of the brain) also separate units? Or did their computational role require them to be interconnected to form a so-called reticulum of continuous cell membranes? The latter idea was widely held, but no one could be sure. The problem was, neurons extend long, thin, and branching biological wires (axons and dendrites) that were hard to stain and nearly impossible to follow under the microscope. To put it simply, there was no

FIGURE 3.1. One of Cajal's childhood drawings of a laborer in a tavern, drawn from memory and with no formal instruction in art. He drew it at age nine or ten, shortly after he was told he had no talent.

way to see the full extent of neuronal trees (neurons) because of the dense forest of intertwined and closely associated branches. All that changed when an Italian histologist by the name of Camillo Golgi invented a new stain (aptly called the Golgi stain).

Golgi's stain takes advantage of the fact that under the right conditions certain silver compounds (silver nitrate) have an affinity for the cell membranes of neurons. When properly prepared, blocks of neuronal tissue can be "developed," or made selectively dark, much the way silver was once commonly used to darken film or paper in photographic darkrooms. But there is one very mysterious and special property of Golgi's stain that we still do not understand—it only darkens a small percentage of the neurons. Those few neurons are made dark in their entirety, down to the last small branches of the dendrites and axons, while the rest remain clear. In other words, a few trees stand out clearly from the tangled forest.

Golgi published the first results of his stain in the early 1870s,[4] but, strangely, many years passed before other histologists understood what Golgi had invented, in part because petty rivalries discouraged many scientists from adopting someone else's invention. And yet, when Cajal was shown a Golgi-stained neuron by a friend in 1887, he was an immediate convert.[5] He knew the stain was key to solving one of the great puzzles of nature—namely, what is the fundamental unit of the brain? And that's not all; he thought Golgi-stained neurons were beautiful, as best described in his own words:

The garden of neurology holds out to the investigator captivating spectacles and incomparable artistic emotions. In it, my aesthetic instincts found full satisfaction at last. Like the entomologist in pursuit of brightly colored butterflies, my attention hunted, in the flower garden of the gray matter, cells with delicate and elegant forms, the mysterious butterflies of the soul, the beating of whose wings may some day—who knows?—clarify the secret of mental life.[6]

With the new stain in hand, Cajal set forth on one of the most impressive scientific marathons in history, working tirelessly to document the structure of neurons and even brain circuits in countless species at different developmental stages. By producing literally thousands of elegant and minutely accurate illustrations, Cajal convincingly demonstrated that the brain is composed of discrete cells, a conclusion known as the neuron doctrine. He also correctly deduced the direction of information flow in neurons (from bushy dendrite to single axon—see figure 3.2).[7] With the latter conclusion in hand, he was able to also (correctly) propose how information flowed within entire circuits, for example, in the human spinal cord and retina. He also discovered and named the "growth cone," the means by which developing neurons extend portions of their cell membrane to make connections with other neurons, and suggested (correctly) that growth cones find their targets by following chemical gradients produced by other cells. It is hard to overemphasize Cajal's impact on the field of neuroscience, and it is astounding that he did much of his work in the 1800s. Not surprisingly, he was awarded the Nobel Prize (in 1906).

Having given you the short punchline from his life's work, I want to turn back to how Cajal achieved such success. The answer is not as simple as it seems, because Golgi—the very inventor of the stain that Cajal used to such great effect—was vehemently opposed to Cajal's conclusions. Golgi believed that neuronal circuits were not composed of discrete cellular units, but were instead contiguous, joined to one another

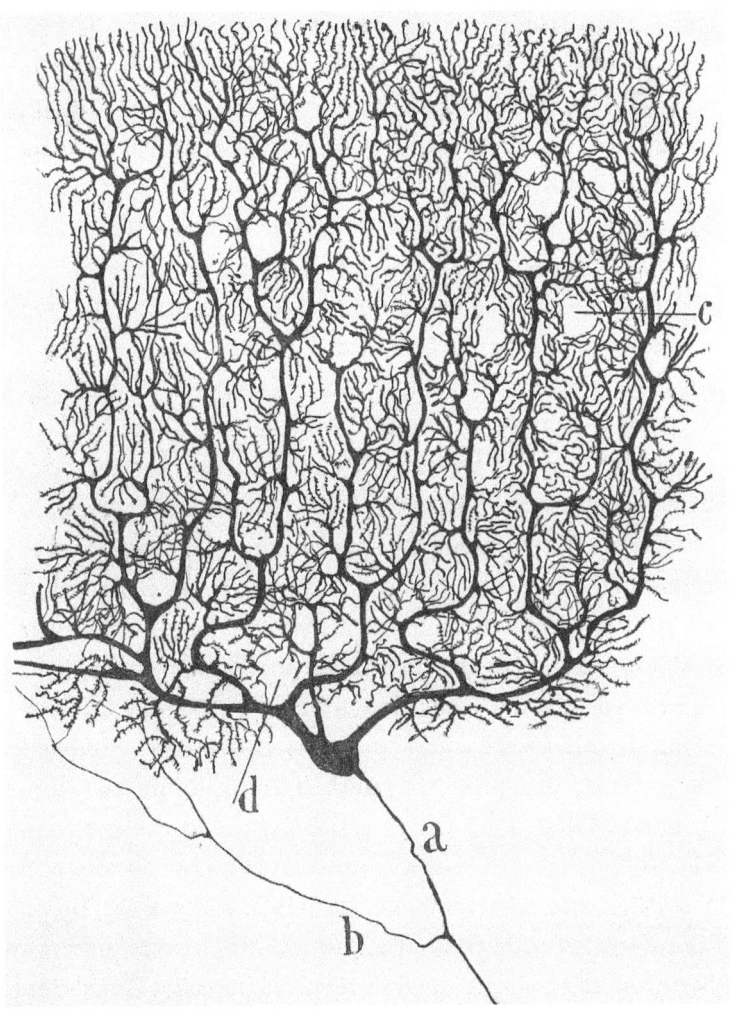

FIGURE 3.2. A neuron (Purkinje cell) drawn by Cajal from the cerebellum of a bird.

rather than merely making close contact. This was called the reticular theory, and it had many supporters. In fact, Golgi shared the 1906 Nobel Prize with Cajal, and never gave up his misguided belief in the reticular theory. That said, it is a great historical irony that Cajal's skillful use of Golgi's stain ultimately won the day, as Golgi's supporters progressively dwindled in the face of Cajal's prodigious scientific onslaught.

In his autobiography, Cajal posed the question of his own success, asking: "Why did my work suddenly acquire surprising originality and broad importance?" In answer, Cajal highlights two key and very practical contributions.[8] First, through long and careful experimentation he made substantial improvements to Golgi's stain. Second, he turned the stain toward developing animals and embryos, which showed neurons and their connections with exceptional clarity. As he put it, "As a crowning piece of good fortune, the chrome silver reaction, which is so incomplete and uncertain in adults, gives in embryos splendid colorations, singularly extensive and constant." In other words, Cajal took up where Golgi left off and then revised, extended, and polished the experiments.

Finally, Cajal took artistic inspiration from his preparations. He sought and found beauty in the structure of neurons, and translated that beauty into incomparable works of art. This is no exaggeration; drawings of Cajal's neurons have been exhibited around the world, and they remain a staple for illustration of introductory textbooks in neurosciences to this day. They are accurate and beautiful representations of neuron structure that have seldom been surpassed.

But I started out this chapter by saying there was more than aesthetics at stake when it comes to art in science. At the risk of being decidedly unromantic about artistic inspiration, I want to turn to the more practical and strategic lesson that can be drawn from Cajal's search for aesthetic landscapes in the brain. And, for now anyway, I'm not talking about persuasion of other scientists.

Consider the lesson of chapter 1, that having a scientific puzzle is a means to focus and maintain your attention. Cajal certainly had a grand puzzle in hand: What is the fundamental unit of the brain? In chapter 2 I further suggested that revising experiments is another means of maintaining and honing your attention. Cajal attributes a major part of his success to first improving the Golgi stain and then finding the very best place to use it—behind that statement lie countless views of diverse neurons in a range of different animal species.

And now I come to the great utility of the search for beauty and inspiration in data—it is another powerful means of focusing and holding your attention. You cannot, for example, draw every minute detail of a complex three-dimensional neuron, and repeat the process hundreds of times, without being intimately close to your data. When all of these motivations— puzzle-solving, revising experiments, plus searching for and documenting beauty—combine, you have multiple forces adding together to powerfully focus attention for long periods, which is a sure recipe for discovery. Cajal's obsessive focus on neurons is legendary—he often spent fifteen hours a day at the microscope and produced a veritable flood of publications. As for discoveries, there are too many to list, though you have heard some of the highlights.

Cajal is, of course, an extreme example. I am not advocating the life of a scientist-monk. But I *am* suggesting that part of

Cajal's success was derived from the pleasure he took from the beauty of cellular structures. And while Cajal was an exceptional artist, you need not have innate artistic abilities to find and document artistic landscapes in your data (I speak from experience, as you will see). Moreover, an aesthetic view of your study system is not restricted to "visual arts"; it carries over to all kinds of other data as well. Many neurophysiologists would tell you the same thing about the responses from a great cell while making a chart-recording from the brain, for example. In any case, let me now turn from Cajal to the now familiar example of jewel wasps that I introduced in chapter 1.

Beauty and the Beast

Recall that I brought jewel wasps to my lab without any particular research agenda. How would I interact with this creature, so foreign to us in both anatomy and behavior? Where would I even start? Photography, of course.

I especially enjoyed the challenge of finding art in the case of jewel wasps. After all, the wasp's life cycle is notoriously creepy, and it includes a much-maligned insect (the cockroach), ending with a grisly, eat-your-host-alive finale. It's not exactly the kind of biology you would consider aesthetic. That said, the jewel wasp got its name for a reason—it is a particularly elegant creature as insects go, with a shiny metallic cuticle and the agility of a fencer. Consider figure 3.3. We can easily read what's about to happen from, I dare say, the "expressions" of the cockroach and the wasp.

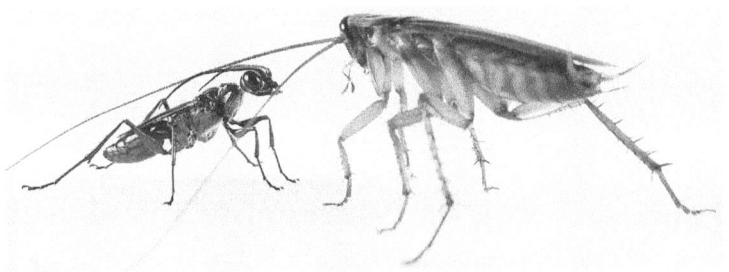

FIGURE 3.3. An emerald jewel wasp stalks an attentive American cockroach.

Both are clearly paying rapt attention to their adversary. The wasp is poised to launch an attack, with its normally forward-facing antennae pulled back over its head, a posture anyone who has worked with horses (or had a pet cat) would recognize as aggressive—for the shared purpose of protecting vulnerable sensors on their head during a fight. For its own part, the cockroach is raised up in a defensive posture called stilt-standing, making itself look larger, very much like a frightened cat with raised back and hackles up.

You can perhaps tell from my description that the process of photographing these creatures taught me a lot about their behavior. To capture the fast action and keep everything in focus, I had to learn, as far as possible, what each insect's behavior might predict about their future moves.

My own next move was to use the scanning electron microscope to get a different kind of portrait. This instrument can go from 10 times magnification to 100,000 times with the twist of a dial. Using this microscope on an insect is like taking a helicopter tour of the Grand Canyon. I began with a portrait (at

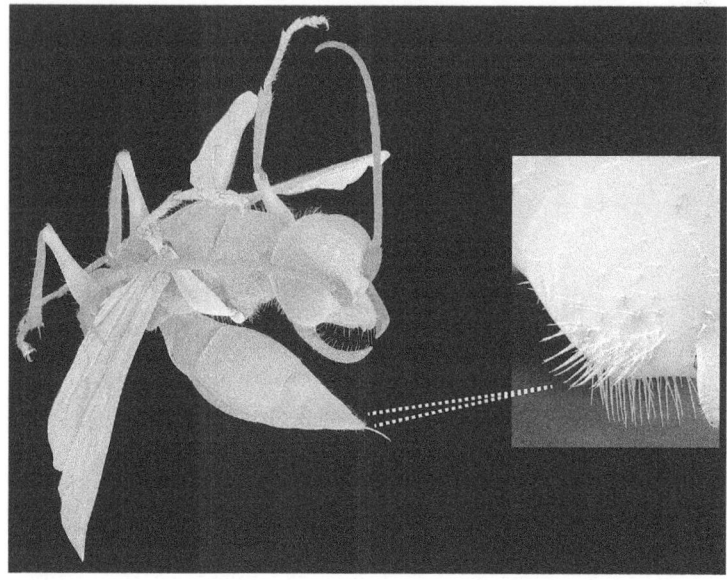

FIGURE 3.4. At left, a full-body portrait of an emerald jewel wasp under the scanning electron microscope, artistically posed for dramatic effect. At right, a close-up of the sensory hairs at the tip of the abdomen.

left in figure 3.4), but then I continued my tour, and that's when I noticed the hairs at the very tip of the wasp's abdomen.

And now I have come full circle to the puzzle described in chapter 1. The search for compelling views of this enigmatic insect played an important role in helping me recognize the puzzle of the sensory hairs that are used by the wasp to find the best place to lay an egg. The jewel wasp example is no fluke—next I'll briefly describe how the search for an aesthetic view led to discoveries in a much different animal, the star-nosed mole (*Condylura cristata*).

From Nose to Brain

Why does a star-nosed mole have a series of bizarre-looking appendages that surround its face? This question was the focus of many of my early studies. Here I will tell one short part of that long story because it all started with my attempt to get some good images of the mole's star under the scanning electron microscope.

Although biologists had wondered about the function of this strange animal's nose since it was first described in the 1800s, few studies had focused on this species because star-nosed moles are very difficult to find. They are not endangered, but they are reclusive and live in the muddy soil of overgrown wetlands. Fortunately, I was able to find them, having worked for the National Zoo in Washington, DC, where part of my job as an undergraduate was collecting and studying star-nosed moles. Later, when I was in graduate school, I decided to take up the mystery of the star once again. I eventually turned to the scanning electron microscope, reasoning that important clues to the star's function might come from a very detailed view of its external anatomy. But that's not all that I had in mind.

The main strength of an electron microscope is *very* high magnification. Cajal's microscope probably topped out at around 1,000 times magnification, whereas the electron microscope can magnify a structure 100,000 times or more. And, as a logical corollary, as you look at tissue at higher and higher magnification, you typically look at smaller and smaller samples. Unless, that is, you are shooting for art.

For an aesthetic view of the mole's star, I began by setting the microscope to a piddling ten times magnification. In fact,

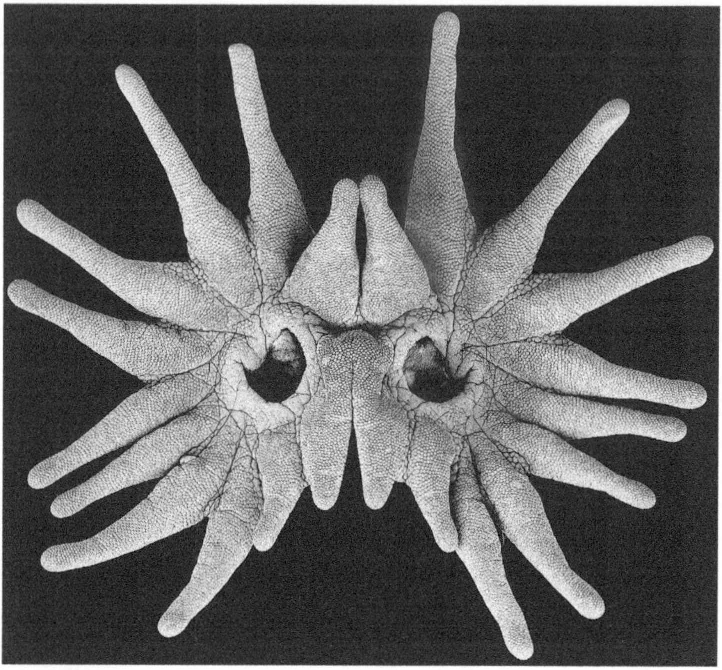

FIGURE 3.5. The mole's star at low magnification under the scanning electron microscope.

I had to push the limits of the microscope in an unconventional direction—pushing it to the very lowest magnification possible. I was rewarded with the image in figure 3.5. It didn't take long to go from that image to key data; by increasing the magnification I soon learned the star is covered entirely with thousands of tiny epidermal sensory organs called Eimer's organs. These are visible as a honeycomb pattern of little swellings on the skin surface of each appendage. Eimer's organs are

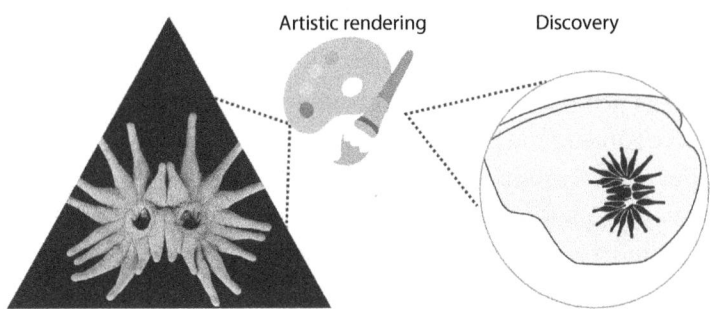

Artistic rendering   Discovery

FIGURE 3.6. Using low magnification to image the star, with the goal of getting an aesthetic image, eventually led to the discovery of the mole's unusual brain map.

exquisitely sensitive to the slightest contact, giving star-nosed moles an unparalleled sense of touch.

But my low-magnification, artistic approach to the star also gave me a different, subtle, and very important insight (reminiscent of Cajal's insight about neurons). I learned that each appendage of the star is a separate sensory "unit" such that the sheet of sensory Eimer's organs that covers one appendage's surface is always separated from the sensory organs of its neighbor by a thin border. This made me wonder about how the star might be represented in the mole's brain. I knew that in some species, you can actually see a reflection of separate body parts in a so-called brain map of the sensory surface. But for this to happen, the touch sensors have to have distinct borders on the skin surface—as I had just observed for the Eimer's organs on the mole's star. To make a long story short, further experiments revealed a "visible brain map" in the star-nosed mole (see figure 3.6).[9] And, as so often happens in science, that finding, in turn, raised many more puzzles and questions leading to more research.

I gave these two examples from my own research to reinforce some of the practical benefits of adding artistic inspiration to the collection of data. In my own case there is almost always a synergy between scientific puzzles and art. Having said that, I can imagine a new student to science thinking: "Oh great, isn't it hard enough to collect decent data? Surely it's twice as hard to aim for art at the same time?" Actually, it is the complete opposite. Most of the battle in science is finding a way to become enthralled.

Flipping the Gold Coin

So far, I have talked about artistic inspiration from an unconventional perspective—for its value to the scientist *at work*. In the next sections (and in the next chapter) I'm going to flip to the more conventional perspective, changing the view from the experimental and discovery side of science to the presentation side of science. After all, art is meant to be shared, as are scientific results. Cajal did not make every neuron he drew into a masterpiece simply to decorate his laboratory. He well knew the value of compelling art for communicating his science.

Here again we can take lessons from professional writers, and this time the lessons are much more direct. For many forms of writing, whether explicitly stated or not, the beginning—the so-called lead—is key to holding the reader's interest. As William Zinsser (the renowned author of *On Writing Well*) puts it: "The lead must grab the reader with a provocative idea and continue with each paragraph to hold him or her in a tight grip, gradually adding information." The same

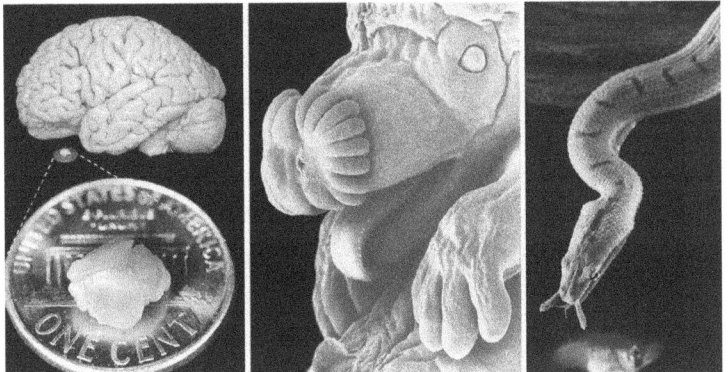

FIGURE 3.7. Three introductory images from different publications by the author.

can be said for images. You might object that a scientific report will not have the gravitas of a front-page article in the *New York Times*. But isn't that all the more reason to try to engage your audience from the start?

And speaking of newspapers, there's a long history of research into what grabs readers, the so-called entry point for their attention.[10] It probably won't surprise you to learn that a large, arresting image gets the most attention. Accordingly, if at all possible I like to start my papers with a simple image that draws the reader in. Ideally the image encapsulates some component of the research that helps to carry readers further into the paper. Three examples are provided in figure 3.7, from three very different kinds of study.

On the left is an image introducing readers to our studies of the shrew brain.[11] Why should anyone care about shrew brains? As it turns out, shrews have among the smallest brains of all mammals, which means they have about the smallest neocortex. Our large and convoluted neocortex (left, top) gives us impressive cognitive abilities, but we evolved from shrew-like ancestors. What were these tiny brains like? Shrews give us our best view of this early state, and the image encapsulates the comparison made in the study, while emphasizing the extreme change in size that has occurred in the course of mammal evolution. With these ideas in mind, readers are (hopefully) more likely to work their way through the detailed aspects of brain mapping and comparative neuroanatomy that follow.

The image in the middle is an embryonic star-nosed mole under the scanning electron microscope. It introduces our work on the relationship between the development of the mole's star and the development of the mole's brain.[12] The image is so bizarre it grabs your attention no matter the context. That said, the developmental theme of the study is suggested by showing an embryo, and part of the main finding from the research is hidden in plain sight—the earliest part of the mole's nose to develop (the front bottom) is also the earliest to develop in the corresponding brain map for the sense of touch.

Finally, the image on the right is simply a photograph of an aquatic tentacled snake. The portrait shows off the snake's unique appendages, used to detect water movements generated by fish. The fish lurking in the background foreshadows the predator-prey interaction.

Those images are examples of leads, used to hook the reader. But don't forget the rest of Zinsser's advice—you have to continue to hold them. You can use artistic additions sprinkled along the way for this—I sometimes think of them as gold coins for the reader—little rewards to be picked up as they carve a path through the jungle of detail and data (I've borrowed the term "gold coins" from Don Fry).[13] An example is shown in figure 3.8 and table 3.1, from a study mapping the

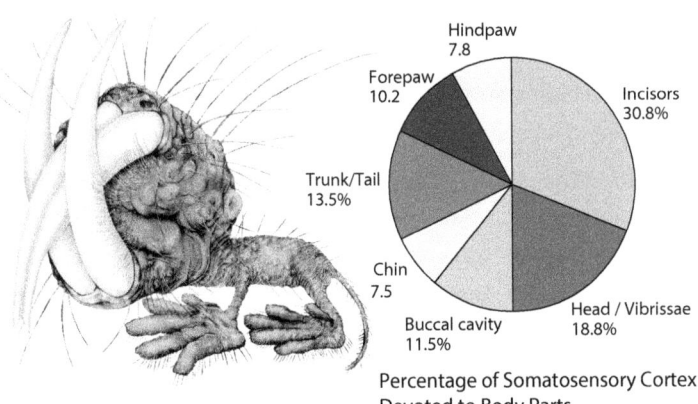

Percentage of Somatosensory Cortex
Devoted to Body Parts

FIGURE 3.8. A "mole-ratunculus" created by the artist Lana Finch to illustrate the relative sizes of body parts mapped into the mole-rat's brain is shown on the left. The sizes were determined by mapping responses in the somatosensory (touch) area of the neocortex. The same data are shown as a more traditional pie chart on the right, and we also provided the raw data for anyone interested in more detail (table 3.1). Images from Catania, K. C., & Remple, M. S. (2002). Somatosensory cortex dominated by the representation of teeth in the naked mole-rat brain. *Proceedings of the National Academy of Sciences, 99*(8) (April 16), 5692–5697.

TABLE 3.1. Proportions of S1 devoted to different body parts

Body part	Case 1 (mm2)	Case 2 (mm2)	Case 3 (mm2)	Case 4 (mm2)	Average (mm2)	St. Dev. (mm2)
Incisors	27.1% (2.9)	30.4% (2.4)	34.7% (4.2)	30.8% (3.5)	30.8% (3.2)	± 3.1% (0.8)
Head / vibrissae	23.3% (2.5)	20.9% (1.7)	14.7% (1.8)	16.0% (1.8)	18.7% (1.9)	± 4.0% (0.4)
Buccal cavity	10.6% (1.1)	11.6% (0.9)	12.9% (1.6)	10.8% (1.2)	11.5% (1.2)	± 1.0% (0.3)
Chin	5.5% (0.6)	10.5% (0.8)	7.3% (0.9)	6.6% (0.8)	7.5% (0.8)	± 2.2% (0.1)
Trunk / tail	12.0% (1.3)	11.3% (0.9)	14.1% (1.7)	16.7% (1.9)	13.5% (1.4)	± 2.4% (0.4)
Forepaw	11.3% (1.2)	10.5% (0.8)	7.7% (0.9)	11.4% (1.3)	10.2% (1.1)	± 1.7% (0.2)
Hindpaw	10.2% (1.1)	4.7% (0.4)	8.7% (1.0)	7.7% (0.9)	7.8% (0.8)	± 2.3% (0.3)
Total S1	(10.6)	(7.9)	(12.0)	(11.4)	(10.5)	± (1.8)
Total neocortex	**(39.8)**	**(25.3)**	**(37.2)**	**(32.5)**	**(33.7)**	**± (6.4)**

brain areas devoted to touch in the neocortex of the naked mole-rat (*Heterocephalus glaber*).[14]

The results were shown in three different ways on a single page of the publication. For the neuroscience connoisseurs, who might be interested in the raw data from each experimental case, a table of values was provided. But these data are much more easily and quickly digested in a pie chart, giving the percentages of brain territory devoted to each body part. The surprising result is that almost a third of the territory is devoted to processing information from the front teeth—the incisors. But the real attention-getter is the cartoon image of a mole-rat body on the left, showing each body part in proportion to its size in the brain map. You might think of this as the brain's view of the body when it comes to the sense of touch. The fanciful image is unconventional for the venue, the *Proceedings of the National Academy of Sciences*, but we added it as a fun way to summarize the results.

Do these gold coins matter? I am reticent to criticize my fellow scientists, so I'll let a Nobel laureate do the dirty work. In his book *Brain and Visual Perception*, David Hubel complains that "reading most papers today is like eating sawdust."[15] I don't feel quite that extreme, but many scientists seem to think making discoveries is their only job and communicating is an afterthought. It is true that you can have a career in science without giving much thought to communicating; after all, scientists usually have a captive audience of fellow investigators who speak their unique language. But most scientists would benefit from having their papers more widely read, and this is especially true at the early career stage.

Many people embrace this sentiment when it comes to writing, but forget that it applies to the artwork as well. The figures and writing should be an integrated whole, meaning the goals of communication and engagement apply to both. And, as with writing, a good starting point is to put yourself in the reader's shoes. When you are trying to make sense of a paper, what do you find annoying, distracting, or confusing? For me, abbreviations are at the top of the list. If a compelling image is a gold coin pulling the reader forward, abbreviations are barbed wire blocking the path. Unfamiliar readers are usually forced to double back, retracing their steps in hopes of finding out where the abbreviation was first defined.

Most abbreviations could be avoided to the benefit of all. My best guess is that the overuse of abbreviations was adopted before journals were online and the printed pages were expensive and limited. Despite the relaxed constraints on space that came with the digital revolution, the compulsion to abbreviate remains. I hereby dissent.

Although some abbreviations have their place, I suggest avoiding all but the most commonly known abbreviations in your field (unless space is truly a constraint for your publication). But why stop there? Many scientific illustrations are covered with abbreviations that are defined in the figure legend, when the full text might be easily printed on the figure, saving the reader valuable time and effort. (If you need a scale bar on an image, consider labeling it with the length, which will make it easier to interpret *and* save space in the legend.) Even authors that sensibly meter their use of abbreviations in the text seem to instinctively abbreviate everything on an image.

I can't help but return to David Hubel, as I seem to have converged with his views on this matter. In the epilogue of his book on visual cortex he states:

> Above all we tried to make our papers easy to read. With almost religious determination we avoided abbreviations, with the sole exceptions of LGB and EPSP. That avoidance probably cost the publishers all of three extra lines at the end of some of our papers, but it must have saved our readers many minutes of searching back to find what the letters stood for. We tried to make our illustrations easy to read by resisting the temptation to combine eighteen small figures into one large one, a custom perhaps started by Eccles and perpetuated by Eric Kandel. We tried to avoid figure-legends in which the a's, b's, and c's are buried in the text instead of being separated by paragraphs, so that the poor reader has to search for the tiny letters with a dissecting microscope. We were often unsuccessful because of the puzzling resistance of editors. All this was in attempts to make our papers less tiring and tiresome to read: we hope we succeeded.[16]

If the Nobel Prize is any measure, they succeeded. But I want to return to some of their comments. I'm obviously not providing a detailed manual of how to make figures. Instead I suggest adopting a particular perspective, made clear in the quote above. It has two main facets. First, always think about your reader as you write and as you make illustrations. Second, when you are reading other papers, notice and adopt the things that work well and—equally important—notice and avoid the things that make reading and interpreting papers unnecessarily difficult.

Tools

I was fortunate to experience the pre-digital world, because the labor-intensive process of hand-making illustrations instilled a certain discipline. If the final product wasn't adequate there was no "undo command"; instead, you had to go through the whole process again. So when I bought my first computer (a Macintosh Centris 610) I was ready to appreciate illustration programs for what they are—a time-saving miracle that puts the power to make complex figures into anyone's hands. I can still recall giving my labmate this advice: "If you want to graduate a year early, buy a computer and learn an illustration program."

Although I don't have any statistics, I suspect the second part of my advice may still hold true for many students today. And as for scientists more generally, being able to easily produce, edit, and modify your own illustrations provides the same kinds of benefits (albeit at a different stage in the process) as being able to efficiently modify your experiments (see chapter 2). You can always outsource your illustrations, but you'll have to get in line.

I'll add that with the right strategies, the least artistically inclined can produce quite reasonable artwork with an illustration program—you don't need to be Ramón y Cajal. I speak from experience—it's not false modesty to say I am among the worst freehand artists you could ever meet. My handwriting is equally atrocious, and the only comfort I take

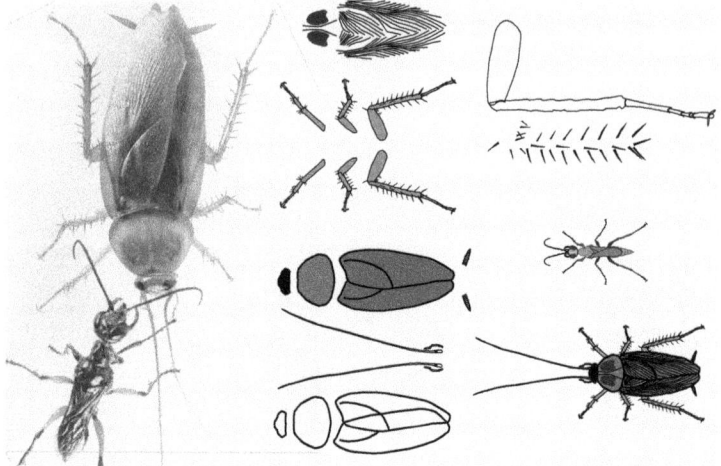

FIGURE 3.9. A breakdown of the simple process for making a reasonably complex rendition of an insect (shown in steps for the roach). I was able to quickly make images of the wasp and roach (using the photo as a rough guide) and use them in various forms in a range of subsequent publications.

from my classroom drawings is that they provide comic relief. Which is to say—if I can do it, anyone can do it. I happened to start with Adobe Illustrator for digital illustrations, and it's one of the few times I actually went through the entire tutorial to learn a program. It has served me well ever since (see figure 3.9).

A second tool, or perhaps I should say skill, that I recommend is photography. Today's students have a leg up in the process, considering almost everyone has a camera in their phone. For most people taking pictures is second nature, now ranking with texting as a ubiquitous practice. But a phone camera is only a start.

It's useful to move up to a more sophisticated camera and learn about the three basic adjustments—shutter speed, aperture, and film speed (which is now simply called ISO). Familiarity with these adjustments is a core skill for photography, and what applies to, say, nature photography also applies in many ways to light microscopy, scanning electron microscopy, confocal microscopy, and many other advanced imaging techniques. Plus photography is fun—how often are you tasked to learn a skill that can be honed by taking a camera to a park, or downtown, or even practiced at home on children and pets?

Finally, I'll take a page—or more accurately a sentence—from Richard Hamming's book of advice for teaching electrical engineers: "Teachers should prepare the student for the student's future, not the teacher's past."[17] In this vein, I'll add that learning how to make and edit videos is a great boon when documenting research results and more generally for science communication. I suspect this trend will only grow in the future.

Value Added

Clear and engaging artwork will probably help your papers pass the review process, and it will almost certainly increase readership among your peers. But if you have a story to tell, there's a whole other group that might want to hear about it—this includes popular science writers, textbook authors, children's book authors, and the popular press. One of the best ways to pique the interest of this diverse group is with compelling imagery.

To give a recent example, in the midst of the COVID-19 pandemic the United States Centers for Disease Control and

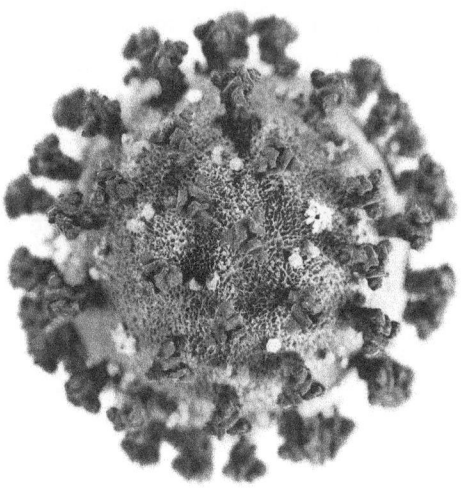

FIGURE 3.10. Black-and-white version of the coronavirus image created by medical illustrators Alissa Eckert and Dan Higgins at the Centers for Disease Control and Prevention.

Prevention was challenged to engage the public with constantly changing information about the virus that causes COVID-19. They asked their medical illustrators, Alissa Eckert and Dan Higgins, to give an identity to the virus— "Something to grab the public's attention."[18] Their mug shot of the virus—an ominous, spiky ball of contagion—became ubiquitous (see figure 3.10).

It showed up in newspapers, behind newscasters, on web pages, in tweets, and pretty much anywhere the virus was being discussed. The image focused the public's attention on one of the most important issues of the time. But it was more than just visual icing—it helped people imagine how masks reduce contagion, and think about why the changing structure of the prominent spike proteins is so important for immunity. Plus it helped people contemplate what might be covering the table or floating in the air at a busy restaurant. It's likely the illustration saved lives.

Most scientific illustrations are not going to grab so much attention, or touch on such a weighty topic. On the other hand, isn't that all the more reason to make them as compelling as possible? News organizations and science writers are interested in all kinds of science stories, and in my experience the amount of coverage increases in proportion to the types of artwork you can provide. A good science story is the minimum (see chapter 4). A good story with some striking imagery will generate a lot more interest and will travel farther. But you'll get the most mileage with a good story, some striking imagery, and a video.

You may not consider popular science communication to be important. Some scientists prefer to think of their job as restricted to conducting experiments and reporting discoveries in technical journals. For my own part, I enjoy telling the world about new findings if I get the chance. Part of this stems from the influence children's books and magazines had on me growing up, without which I might not be a scientist. The National Science Foundation also encourages scientists to communicate their discoveries to the general public, and so the chance of getting funded is better if you are engaged in these so-called "broader impact" activities. That said, it would be disingenuous of me not to admit that getting attention for the work is fun. And some of the most fun comes from getting one of your images on the cover of your favorite science journal.

The Rolling Stone

Musicians dream about being featured in *Rolling Stone* magazine or, better yet, being on the cover, as the song goes. Scientists seldom sing about it, but they, too, are eager to get their work on the cover of a journal (the walls of many labs are decorated with these small triumphs). And there are plenty of opportunities. Even as the landscape changes from hard-copy to online publications, editors still look for that hook, that visual lead that will grab a reader's attention. The competition to be on the cover seems to parallel journal selectivity and impact factor. Some journals don't get enough cover suggestions to easily fill each slot; others have plenty of submissions to choose from, so the competition is stiff. The most prestigious journals have their own art department to design in-house covers, or they commission imagery from professional artists and photographers, rarely considering author submissions.

No matter the case, you're unlikely to get your work on the cover if you don't submit a suggestion. The range of imagery that might be considered is immense—if you happen to be an astronomer, you've got the advantage. But the life sciences are also full of striking imagery, ranging from habitats, animal portraits, fluorescently labeled cells and organelles, scanning electron micrographs, all the way to structural biology (not to mention viruses). My own work often includes animal behavior, and I've had good luck with portraits. Figure 3.11 shows some examples (the crocodile image was taken by a graduate student, Duncan Leitch).

You may have noticed that one of the images (showing the water shrew catching a crayfish) is from *Natural History*

FIGURE 3.11. Examples of cover images from the author's publications.

magazine, and not from a technical (peer-reviewed) science journal. I had always wanted to publish an article in *Natural History*, having read it so often—especially the columns written by Stephen Jay Gould (now available as a series of books). But how do you do this?

My work on animal senses and brain evolution seemed to be in their wheelhouse, so I hoped one day I might hear from the editors. That day never came. Instead I learned an important lesson while reading John Shaw's books on nature photography. He gives a lot of advice on how to take ("make" in his words) better pictures, but he also gives business advice to photographers. I learned that if you want to publish in these venues, you don't wait for a knock on your door. Instead, it's best to submit a formal article proposal to the editors (that's how I did it).

But John Shaw said something else that caught my attention: "I'm firmly convinced that the very best way to break into nature photography is to write an article to accompany your photographs."[19] It dawned on me that photographers have the inverse challenge of most scientists. By that I mean most scientists benefit from improving their artwork when submitting their "science story," whereas photographers benefit from having a story to go along with their artwork.

Of course, you have to have a "science story" to tell in order to write for a popular audience—and the same is often true for peer-reviewed journals. What if you don't know how your research project and experimental results fit into the bigger picture? In the next chapter I'll talk about story.

4

Story

If you're a storyteller, find a good story and tell it.

—HOWARD HAWKS[1]

DID YOU KNOW that your thumb has only two bones, but the rest of your fingers all have three? I didn't know this fact about my own thumb, with which I am quite familiar and presently using, until I looked it up. But, strange as it may seem, I did happen to know that the panda's thumb is not one of the five fingers of its hand. Rather, it is a functional *sixth* digit that came about as an evolutionary contrivance, constructed from a wrist bone (see figure 4.1). It's one of the most famous stories in evolutionary biology, made popular by the late Stephen Jay Gould.[2] Gould did not discover this remarkable bit of anatomy, but he knew a good story when he saw one, so he wrote about it in his monthly column in *Natural History*, and later titled his well-known book, containing the same essay, *The Panda's Thumb*.

For those not acquainted with Gould's essay, the general lesson is that past history often constrains evolution. In this

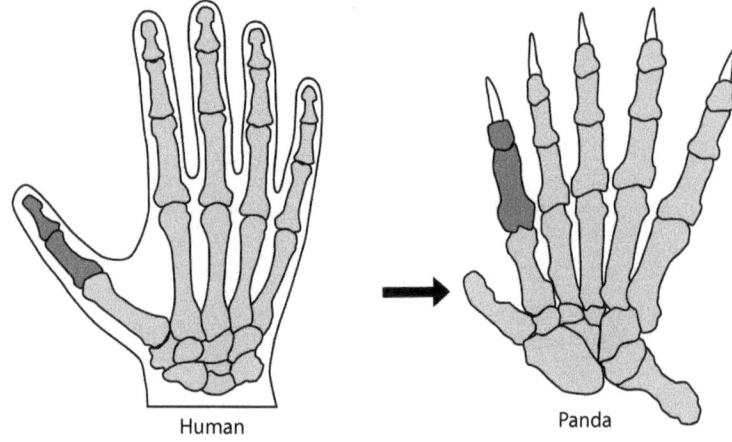

FIGURE 4.1. The equivalent finger bones (phalanges) of a human hand and panda forelimb. Notice the panda's "functional thumb" is not formed from the same bones as typically form the thumb.

case, when the panda's ancestor began to switch from an omnivorous diet to bamboo, the five digits of its forelimb were endowed with claws and already committed to their role in climbing, digging, breaking things apart, and sometimes fighting. But eating bamboo shoots and leaves is greatly facilitated if you can grasp the stalks. Hence there was selection for, and evolution of, a jury-rigged thumb—built from a wrist bone—as a new means for a bear to hold objects tight.

The panda's thumb is only one of many unusual evolutionary contrivances, but this particular example resonates because of how Gould told the story. It includes a charismatic animal with celebrity status at zoos, combined with an overarching biology lesson—namely, that some of the best evidence for evolution is found not in perfection, but rather in awkward and poorly engineered solutions.

In the time since Gould's book was published, the term "panda's thumb" has been adopted in fields as diverse as economics, law and judicial procedure, and intellectual property rights;[3] in each case the analogy is used to highlight historical constraints on change. And of course the term is frequently used in evolutionary biology, where some have even invoked a "Panda's Thumb Principle."[4] All of this is a testament to the power of a well-told story.

You might be used to thinking of story as the realm of fiction writers who have free rein to grab readers and draw them along with whatever comes to mind. But despite the fewer degrees of freedom when writing true stories, there is nevertheless a rich and rapidly growing scientific literature of previous discoveries and ideas to draw upon. Virtually every new discovery is connected to this vast ecosystem of previous work, and telling the story of those connections is a key part of doing science.

This is true especially if you plan to submit your work to one of the top-tier science journals. In fact, for some time the journal *Nature* required authors, upon submission of a paper, to provide a "100 word summary of the paper's appeal to a popular (non-scientific) audience, written at the level of a good national newspaper." I can think of no better testament to the fact that editors want papers that tell a compelling story. And though the stakes may be highest for the most selective journals, you will face the same challenge when you write the cover letter that accompanies almost any submission for publication.

Of course, peer-reviewed publications are not the only place in science where story enters the picture. Talks at conferences, thesis defenses, university seminars, classroom presentations, and a host of other speaking engagements challenge established scientists and students alike to put their findings into a compelling narrative. The standard fifty-minute seminar presentation (or job talk) is perhaps the opposite of a 100-word summary. And yet the stakes are still high—there are few things more unpleasant than being on-stage and having your talk fail. The opposite is equally true—a well-received talk is not only rewarding, it's often required for career advancement.

The search for story in science is much like the search for true stories in journalism. In the latter case, the two-time Pulitzer Prize–winning journalist Jon Franklin explains his success by emphasizing the absolute necessity of putting a story in the proper context. Finding context is one of Franklin's skills, and I could not miss the irony when he posed the following question for budding journalists: "To pluck an analogy from science, have you ever wondered why it is that some researchers consistently make major discoveries while others fail, again and again, to find anything new and worthwhile?"[5] He likens a journalist's quest for stories to that of a scientist—and a key practice in both endeavors is putting the results of research into the proper context of other facts. Franklin became an expert at this—as all experts do—through practice.

As it happens, my own style of exploratory research into somewhat uncharted biological territory has often left me scratching my head, wondering what some strange new discovery means in the bigger picture. The upshot of this recurring challenge is that I've gotten a lot of practice finding context when it comes time to write up my results for publication. Having gone through this process often, I've noticed some consistent themes for the scientific stories I have told. But it's not just me—I've noticed that other scientists' discoveries often fall into one of these categories. I couldn't help but be reminded of the concept of "story archetypes" sometimes discussed in the realm of classical literature (e.g., rags to riches, overcoming the monster, etc.).

These categories include the following: 1) The Exemplar, 2) Mystery Solved, 3) Juxtaposition, 4) Exception to the Rule, 5) This or That, 6) History, 7) Extending the Frontier, 8) Whimsy and Humor.

I'll state at the outset that the list is by no means exhaustive. For example, my own current research is not aimed at human health—but for many investigators the primary category for context of a discovery might be addressing a specific human disease. You will also probably notice that many of the categories could overlap, leaving the investigator with more than one choice of compelling context. So much the better, it's good to have choices, and it's often possible to include more than one theme, especially in the discussion section of a publication. In the next section I'll describe each category with some historical examples, before giving a specific case from one of my own publications. Finally, at the end of this chapter I'll return to how one goes about doing the detective work that helps you flesh out the story behind a discovery.

The Exemplar

Exemplars, or standout examples of a particular phenomenon or aspect of biology, play a dual role in carrying the story of a research discovery. They are typically unusual cases, and may

FIGURE 4.2. A peacock showing off its extravagant plumage.

even be awe-inspiring. At the same time their particulars represent a telling case of a much more widely applicable principle. The peacock's famous tail (figure 4.2) is an obvious example, having begged explanation to the point of haunting Charles Darwin, who once exclaimed, "The sight of a feather in a peacock's tail, whenever I gaze at it, makes me sick!"[6] Darwin wondered how such an elaborate and seemingly maladaptive trait could possibly evolve through natural selection. His later insight—that sexual selection through female choice may also drive evolution—transformed the peacock's tail from an enigma into a success story. In this single extreme example, one gets the dual effect of both awe and explanation.

The unusual anatomy of the orchid *Angraecum sesquipedale* (see figure 4.3), also called "Darwin's orchid," provides another

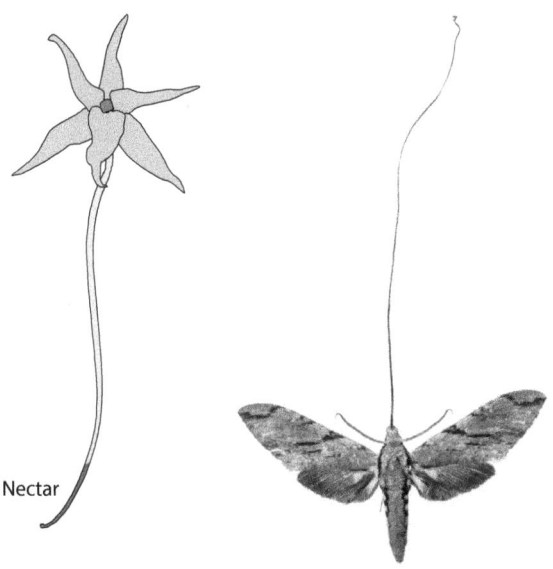

Nectar

FIGURE 4.3. "Darwin's orchid" (at left) has a nectar reservoir that can only be reached by a pollinator with an extraordinarily long tongue. Based on the orchid, Darwin predicted the existence of a long-tongued moth as the pollinator. After Darwin's death, the moth (at right) was discovered.

well-known example.[7] The long extension below the flower, called the "spur," contains nectar that can only be reached by a pollinator with an extraordinarily long tongue. Darwin predicted the existence of a long-tongued moth as the pollinator.[8] Twenty-one years after his death, a moth called *Xanthopan morganii praedicta*, with just such a tongue, was described, and it was later observed visiting the flower as Darwin had anticipated.[9] It is a compelling example of the coevolution of plants and pollinators, and also shows that evolutionary principles allow not only for reconstructing the past but also for predicting future discoveries. (You may have noticed that the panda's thumb is also an exemplar of the tinkering nature of evolution, a process that may result in strange and imperfect contrivances.)

Hearing of these famous examples, two of which hail from Charles Darwin, you might wonder how often this theme could possibly be useful as a context for the findings of a typical scientist. More often than you might guess. I have converged on this theme several times when I realized my results were exemplars of a particular phenomenon or when they highlighted a key trend in biology or neuroscience. You have already read about one example—how tiny shrews provide exemplars of how the neocortex (the outer, six-layered sheet of neural tissue found in all mammals) differs between large-brained mammals (such as humans) and small-brained mammals that resemble our ancestors.[10] The unusual embryonic development of the mole's star[11] is another standout example, fortifying Stephen Jay Gould's suggestion (from his book *Ontogeny and Phylogeny*[12]) that some development sequences may, in fact, recapitulate some parts of evolution.

But my favorite discovery that fell into this category was the exceptional speed with which star-nosed moles can find and eat small prey.[13] It takes a hungry mole only about 200 milliseconds (one-fifth of a second) to touch a tiny invertebrate twice, decide to eat it, and then go on about its way looking for the next morsel of food. The shortest time for a mole to find and eat something small was 120 milliseconds (about one-tenth of a second).

The star-nosed mole is, to put it mildly, blazingly fast—fast enough to earn a place in *Guinness World Records* (see

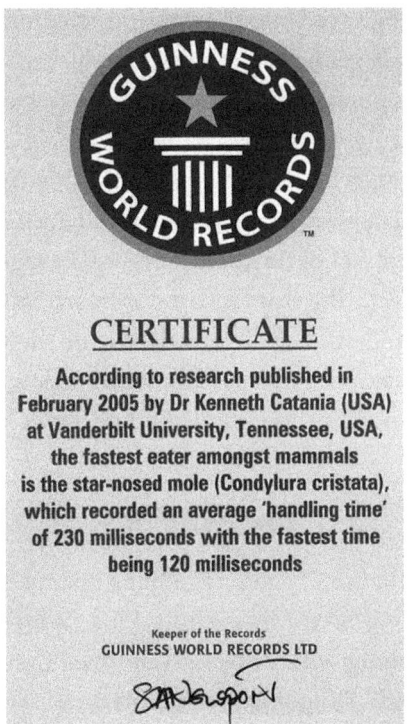

FIGURE 4.4. Official certificate for star-nosed moles as the fastest known foragers among mammals.

figure 4.4). But what's the scientific context that makes the finding interesting in the bigger picture? I didn't know. That is, until my background research took me into an area called "optimal foraging theory."

Optimal foraging theory provides a mathematical analysis of predator behavior based on how long it takes predators to "handle prey" compared with the number of calories (the energy available) in prey.[14] Plugging the star-nosed mole's record-breaking speed into the foraging equations showed the mole could *only* get enough calories to survive on small prey by eating with phenomenal speed. Better yet, many aspects of the mole's biology—small front teeth that act as tweezers, a star covered with touch sensors to find tiny things, special brain areas for high-resolution touch, and even their unusual wetland habitat full of tiny, soft prey—all fit together to tell a compelling story. The story is that star-nosed moles are exemplars of the predictions of optimal foraging theory. This context was a revelation, and it gave me a compelling way to frame an otherwise isolated and quirky finding about fast eating.[15]

Juxtaposition

The *Oxford English Dictionary* defines juxtaposition as "the action of placing two or more things close together or side by side, or one thing with or beside another; the condition of being so placed." I'm going to concentrate on what I consider to be the most compelling opportunity to juxtapose two biological systems—predator-prey interactions, which might be aptly called arms races (see figure 4.5). In my experience teaching about animal brains and behavior, predator-prey interactions are best for keeping students on the edge of their seats. If scientific exemplars can be said to provide both awe and explanation, arms races supply drama and explanation.

One of my favorite examples is what you might call the sensory arms race between bats and moths. In 1938, Harvard undergraduate (at the time) Donald Griffin teamed up with physicist G. W. Pierce and revealed that bats give off pulses of ultrasound as a means to locate large objects and navigate

FIGURE 4.5. Tentacled snake strikes have been pitted against fish escape responses for millions of years.

around obstacles. And yet, at the time, no one could imagine that hunting bats used these same signals to detect and capture tiny flying insects. It took another decade to reveal the full versatility of echolocation as a premier sensory system for aerial predation at night.[16] But even that was only half of the story.

Soon investigators turned their attention to the enigmatic ears found on most species of night-flying moths[17] (a bat's favorite meal). What could these simple ears possibly be used for? They are, of course, used for detecting and avoiding echolocating bats.

FIGURE 4.6. Moths have ears that can detect the ultrasonic echolocation calls of bats, triggering drastic escape maneuvers.

Many moths can hear the bat's echolocating call long before the bat can detect the moth—the sound triggers a range of evasive flight maneuvers that serve to (hopefully) take the moth out of the line of fire (see figure 4.6).

The story has only grown from there. Today we know the sky on warm summer nights resembles a battlefield full of radar-guided predators and radar-detecting prey.[18] There are hundreds of bat species, each using a different hunting strategy, and thousands of moths, many with unique countermeasures. Some moths have sound-absorbing scales thought to provide a form of cloaking from echolocation; others respond to bats with their own radar-jamming ultrasonic countermeasures. Bat echolocation has rightly been called a "magic well," because the more we learn, the more new discoveries come to the surface.[19]

The same coevolutionary drama can play out at the molecular level. Consider the venom of the many-banded krait (*Bungarus multicinctus*). One neurotoxic component of this snake's venom (aptly named bungarotoxin) binds tightly to the molecular channel (the acetylcholine receptor) that is critical for the communication between nerves and muscles. Envenomation results in paralysis, not only of skeletal muscles needed for escape or for putting up a fight, but also of the diaphragm, which is required for respiration. Hence the snake's bite is frequently deadly, even to humans. But there is far more to the story.

A number of mammals (such as the mongoose) have turned the tables on the snake by evolving acetylcholine receptors that are resistant to the venom.[20] This in turn has helped investigators understand which parts of the acetylcholine receptor are most important for communication between nerves and muscles. The back-and-forth between predator and prey in this context supplies both drama and the chance for understanding, in this case at the molecular scale.

Next I want to give an example of how I've used this theme. You've already heard about the tentacled snake and its remarkably sneaky attack strategy. To refresh your memory, the snake waits motionless and hidden, looking like a stick, until a fish swims to just the right spot. Then, just before striking out with its jaws, the snake moves its neck (on the opposite side of the fish from the snake's head), generating a water movement that causes the fish to escape in the wrong direction—toward the snake's strike.

When I made this discovery, the fish escape response, and the underlying neural circuitry that controls it, had already been thoroughly studied for many decades. There are well over one hundred scientific papers on the subject, plus an entire book written about one type of nerve cell (the Mauthner cell) that controls the escape response. By delving into this treasure trove of previous work, I was able to show how, and why, the snake's attack works so well.

Juxtaposing the well-known biology of the fish with the novel finding about tentacled snakes allowed me to tell a complete and satisfying story about predator-prey coevolution.[21] When talking about this research in seminars, meetings, or a class, I spend as much time talking about fish, and the neural underpinnings of their escape, as I do about the tentacled snakes.

There are many opportunities to adopt this strategy of juxtaposition when describing the results of biological studies, especially when there is a coevolutionary context. This theme would naturally apply to predator-prey interactions, courtship and mating, social interactions and language (the latter requires both speaker and listener), parasites and hosts, plants and pollinators, venoms and toxins (the latter have a producer and a target), and myriad other interacting systems.

Exception to the Rule

There are over forty species of cormorants that live around the globe, making a living diving for fish. As you would expect, they are all excellent swimmers with webbed feet, and all of

FIGURE 4.7. The flightless cormorant from the Galápagos Islands (bottom) has reduced wings and a larger body, both of which are thought to improve diving efficiency.

them can fly, with the exception of one outlier species (shown on the bottom in figure 4.7). This species should get your attention—what's the deal with those useless wings?

The flightless cormorant (*Nannopterum harrisi*) is only found as a small population on the Galápagos Islands, where it evolved in the absence of terrestrial predators and surrounded by a relative abundance of food. As with so many other species from the Galápagos, the unique environment, combined with millions of years of isolation, allowed evolution to select for unique traits. Diving is apparently more efficient with a larger body and small wings,[22] hence there was not just relaxed selection for flight, but also positive selection for reduction of wing size and enlargement of body size. It's a standout story that has recently been followed to the molecular level as investigators have been able to determine the genetic basis for small wings.[23]

The flightless cormorant is one of my favorite examples of this theme, but then again nearly every animal from the Galápagos is already famous. To give a more subtle but equally compelling example, consider the study by John Ratcliffe and colleagues in 2003, titled "An Exception to the Rule: Common Vampire Bats Do Not Learn Taste Aversions."[24]

Taste aversions are long-standing aversions to specific tastes that develop as a result of getting sick from food. You may have experienced this unfortunate form of learning yourself, if you've ever eaten food that was "off." What makes these associations special is that learning takes place unconsciously, and it happens despite a long period of time that often elapses between eating and becoming ill. The adaptive value of this special kind of learning is obvious: you don't want to eat toxic substances more than once, even if they tasted good.

The experiments described in the paper show that vampire bats do not exhibit this kind of aversive learning (other bats do). Why? Because vampire bats feed only on blood, and blood from live mammals is *never* toxic to a vampire bat—in contrast to, say, many insects, plants, or carrion eaten by other mammals. Hence there has been no selection for vampire bats to retain this special form of learning; they simply don't need it. After reading this paper, I added the example to my class because it's such a memorable case for why taste aversions do (or don't) evolve.

Next I'll give an example of how I've used the exception-to-the-rule theme. I've already told you about how quickly star-nosed moles can eat. But they have other tricks as well. They are semiaquatic, meaning they often dive into the water in search of food. I wondered whether they might forage differently underwater, so I filmed them close up with a high-speed camera while they hunted. That's when I discovered that star-nosed moles can "sniff" while they swim by exhaling air bubbles (which stay attached to their nostrils) and then reinhaling the same bubbles.[25] Experiments showed they can use this trick to follow underwater scent trails. Better yet, I discovered that water shrews (*Sorex palustris*, another semiaquatic mammal that hunts underwater) can also sniff underwater (see figure 4.8).

Underwater sniffing is a great trick, and its novelty makes for a good story. But the story got better when I found out more about olfaction in other semiaquatic mammals (such as sea lions, seals, and whales). Here's an example of what people have said about mammals using smell underwater (from a book on the subject[26]): "It is strikingly apparent that a mammal which seeks its food exclusively beneath the surface of the water . . . has virtually no need for a sense of smell."

By learning about this preconception—that mammals supposedly cannot smell underwater—I was able to cast my new finding not only as an interesting new behavior, but also as an exception to the rule about mammal smell.[27]

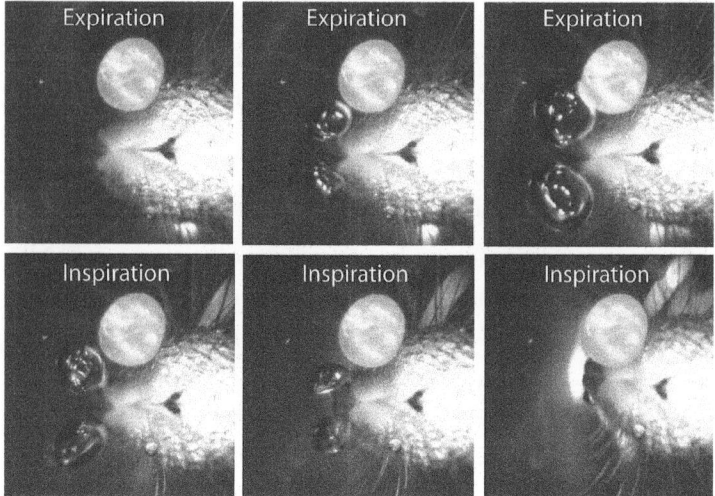

FIGURE 4.8. Images from slow-motion video showing a water shrew, filmed from below, performing an underwater "sniff" by exhaling and then inhaling an air bubble to collect odorants.

Mystery Solved

There are endless mysteries in science. What is the structure of DNA? Is the universe expanding? Is there life on Mars? How did life begin? What is dark matter? Why did the dinosaurs go extinct? People are especially curious about unsolved mysteries, and scientists are no exception.

To stick with dinosaurs for a moment, you are probably aware of the overwhelming evidence that a large asteroid hit the earth 66 million years ago, causing an environmental catastrophe and mass extinction that wiped out three-quarters of life on Earth—including the dinosaurs. Evidence for this event came from an unusual rock layer containing high concentrations of the element iridium, which is rare in the earth's crust but relatively abundant in asteroids. The results were published in *Science* in 1980 by a team that included Luis Alvarez (a Nobel laureate in physics) and his son, Walter Alvarez, along with chemists Frank Asaro and Helen Michel.[28]

When describing their motivation to investigate the unusual rock layer, Luis Alvarez had this to say: "Why did we study this problem in the first place? . . . a few years ago, the four of us suddenly realized that we had combined in one group a wide range of scientific capabilities, and that we could use these to shed some light on what was really one of the greatest mysteries in science—the sudden extinction of the dinosaurs."[29] In other words, as with any mystery, the context for their results existed well before the work began.

Mysteries need not involve catastrophic collisions and planetwide extinctions to be compelling. Consider the epic mystery of freshwater eel reproduction. (In this case, I refer to the true, and very common, freshwater eels in the order Anguilliformes, not the unrelated electric eels from the Amazon.) Since at least the time of Aristotle (born 384 BCE), people had wondered where freshwater eels reproduce, and it's hard to overstate the obsession the scientific community had with this problem in the early 1900s.[30] The question was the biological equivalent of today's quandary with "dark matter" in physics. Many scientists took up the challenge and failed— perhaps most famously Sigmund Freud.

The mystery was finally solved by a Danish investigator, Johannes Schmidt, who spent more than a decade searching for the eel's breeding ground and, by collecting smaller and smaller eel larvae, homed in on the Sargasso Sea in the middle of the Atlantic Ocean. The research was a Herculean task, as best described in Schmidt's own words: "I had little idea, at the time, of the extraordinary difficulties which the task was to present, both in regard to procuring the most necessary observations and in respect of their interpretation. Our work on these eel investigations has now extended some 16 or 17 years. . . . The task was found to grow in extent, year by year, to a degree we had never dreamed of."[31]

His results were published in the prestigious *Philosophical Transactions of the Royal Society of London* and earned Schmidt the Darwin Medal. The story of this scientific quest, in which Schmidt is just one of many characters, is the subject of a delightful book, *The Book of Eels*, by Patrik Svensson, published in 2019.

In 2008 I tried my own hand at solving a fascinating mystery—how does worm grunting work? Worm grunting is a bait-collecting technique that has been used for generations to harvest earthworms in the southeastern United States (see figure 4.9). It consists of pounding a wooden stake into the ground and then rubbing the stake with a flat strip of metal (often an automobile leaf spring) to generate strong, low-frequency vibrations. In response, hundreds of large earthworms come to the surface and are easily collected.

Why this technique of "worm charming" works had been the subject of speculation for over a century—even Darwin wondered why worms come to the surface in response to vibrations.[32] Through a series of experiments in the lab and in the field,[33] I was able to show that the vibrations created by worm grunters simulate digging moles (a worm's archenemy)

FIGURE 4.9. Gary Revell shows off an earthworm captured while worm grunting at the Sopchoppy Worm Gruntin' Festival.

and that the worms collected by grunters have a pronounced escape response from moles. Because moles don't come to the surface to chase worms, exiting the soil when vibrations are detected is a great strategy for worms, *most* of the time. It's the unlucky worms that are tricked by the rare worm grunters and then eventually end up on a fishhook (there are far more moles around than bait-collecting worm grunters). This example is reminiscent of tentacled snakes that trick fish into swimming in the wrong direction, except in this case the sneaky predator is a human.

As with the other studies described above, the context for writing up these results existed long before the study was conducted. Which is to say, interest in a solution to the mystery was guaranteed from the outset. I published the results in the online science journal *PLoS ONE*, and despite the seemingly obscure subject of worm behavior, the paper has been viewed more than fifty thousand times. That's a testament to the enduring interest in mysteries solved.

This or That

If there's one thing more compelling than a scientific mystery, it's a disagreement between scientists who have different answers to the same question. Does the earth orbit the sun or vice versa? Is there a luminous aether that carries light waves? Do the earth's continents move? These and many other questions have been the subject of intense disagreement, and new evidence in favor of one position or another has always been inherently interesting.

Consider the epic debate in neuroscience, between Cajal and Golgi, about whether neurons are separate cells. Cajal had it right, of course; neurons are separate entities. But that, in turn, led to another epic debate—and another Nobel Prize. How is information passed from one neuron to another (or from a neuron to a muscle) where the cells come close together, at the synapse?

In the early 1900s there were two schools of thought.[34] One idea was that communication across the synapse must be electrical, such that a nerve impulse (an action potential) simply passed from one neuron to the next. The other idea was that communication across the synapse was chemical, such that, when the action potential arrived at the synapse from the first cell, some substance was released (a neurotransmitter), crossed to the second cell, and triggered a subsequent action potential (much of the research in this context was conducted at the synapse between a nerve and a muscle).

In 1921, Otto Loewi performed a simple but pivotal experiment (the idea came to him in a dream) (see figure 4.10).[35] He isolated two hearts dissected from frogs, and stimulated the vagus nerve leading to one of the hearts. Stimulation of the vagus nerve causes the rate of the heartbeat to slow. The question was, how? To test the idea that communication from the nerve to the heart was chemical, mediated by some sort of neurotransmitter, Loewi collected fluid from the first (stimulated) heart and applied it to the second (unstimulated) heart. As a result, the second heart slowed its rate as well.[36] This simple experiment provided very strong evidence that some form of chemical transmission occurred between the vagus

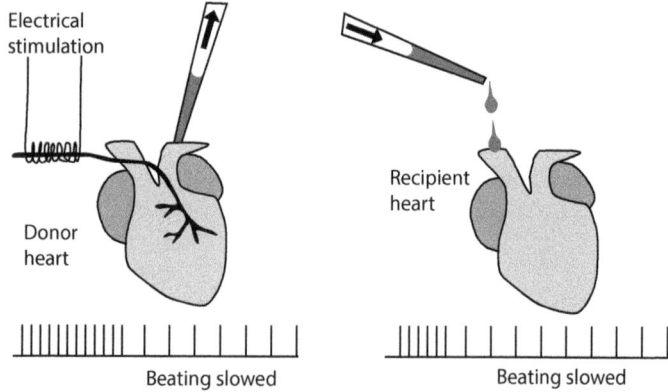

FIGURE 4.10. A summary of Otto Loewi's pivotal experiment demonstrating the chemical nature of synaptic transmission. After the (vagus) nerve leading to one heart was stimulated, and the heartbeat slowed, Loewi transferred fluid from the first heart (at left) to a second heart (at right), and its rate of beating also slowed. This confirmed the existence of a chemical signal, which Loewi called "vagusstuff." We now know that vagusstuff is the neurotransmitter acetylcholine.

nerve and the heart. (In this case, a neurotransmitter, later identified as acetylcholine, had an inhibitory effect.) Otto Loewi's experiment earned him a Nobel Prize and led the way to the acceptance of chemical neurotransmission, which is the predominant mechanism by which signals are passed from neuron to neuron, and from neuron to muscle.

I especially like to teach about this history, because we can all understand why it was hard to believe neurotransmission at the synapse could be chemical. After all, it takes only about half a millisecond (1/2000th of a second) for the chemical signal to be released, diffuse to the next cell, and activate the receptors that generate the electric signal in the second cell. This *seems* impossibly fast, as opponents of the idea would often argue.[37] Which goes to show that our imaginations can often be the limiting factor when it comes to accepting reality.

Next I'll turn to a debate about the fundamental organization of the neocortex in mammals that I happened upon in the course of my own research. My part in this debate started when I discovered that star-nosed moles have a very unusual brain, in which you can actually see (when the tissue is properly stained) an anatomically visible map of the star in the touch area of the neocortex. The practical upshot of the discovery was my ability to make very precise measurements of the areas of the neocortex devoted to each part of the star.

I noticed that one part of the "star map" was much larger than you would expect from the corresponding part of the star on the mole's nose. Why would this be the case? It turned out that this central part of the star acts like a touch fovea—used by the mole for detailed investigations, the same way we use the center part of our retina to scrutinize visual scenes in detail.[38]

I decided to look for other similarities between mammal visual systems and the mole's touch system. That's when I learned about an ongoing debate about how the neocortex is organized.[39] Some investigators thought that the size of each part of a cortical map was determined by the number of nerve fibers coming from the sensors represented in the map (either from the retina in the case of the visual system, or from the star in the case of its touch system). Other investigators thought that extra cortex could be devoted to the most important nerve fibers coming from the sensors. There was debate and

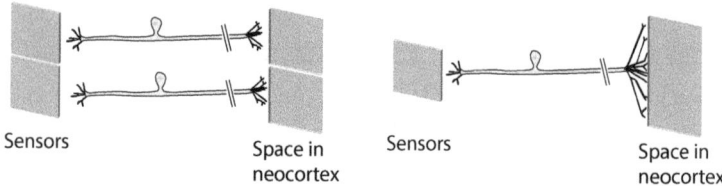

Sensors

Space in neocortex

Sensors

Space in neocortex

FIGURE 4.11. Schematic illustrating different relationships between the sensors on the skin and brain maps in the neocortex. Some studies suggested that the average amount of neocortex devoted to processing information from the skin was always proportional to the number of nerve fibers. But results from star-nosed moles show that some nerve fibers get "extra space" in the neocortex (at right).

uncertainty because it is very hard to measure the details of most cortical maps that are not anatomically visible.

I had the perfect system for answering the question (at least for one species). I had already made very precise measurements of the mole's star map in the neocortex, so all I had to do next was count the nerve fibers coming from each part of the star. My results clearly supported only one side of the debate—the sensory inputs from the important, center part of the mole's star got far more than their fair share of the neocortical map (see figure 4.11). When I wrote up the paper, I was able to introduce my results with the phrase "It has long been debated whether . . ." before showing a definitive answer.[40]

History

So far I've mentioned many historical examples of the different themes I've been suggesting, but sometimes history itself is the theme, or at least it's a big part of it. For example, much of the research that was conducted to find and observe the long-tongued moth that pollinates Darwin's orchid was motivated by the moth's great historical significance. That's the first part of the story—solving any problem posed by Darwin is inherently interesting. But the second part of the story, in the case of the moth, is that subsequent studies also taught us new things about coevolution and pollination. For example, it turns out that predators hiding on flowers may play an important role in the evolution of long tongues—having a longer tongue allows moths to suck nectar at a safer distance.[41]

Another of my favorite examples of the synergy between modern research and compelling history comes from the work of Peter and Rosemary Grant on the Galápagos island called Daphne Major.[42] There the Grants, along with their students, spent many years studying several small populations of Darwin's finches. During an intense drought that occurred in 1977, seeds became scarce and many of the medium ground-finches (*Geospiza fortis*) died. Those that survived had the largest beaks, allowing them to open the scarce, large seeds that remained. When these birds later reproduced, their descendants had (on average) larger beaks than the predrought population. It was an astounding result, showing that evolution, even in vertebrates, can happen in only a few short years.

The Grants' research has all the elements of a great science story. The findings are important and compelling, and the study system is historically fascinating, not only for its interesting geographic location in the Galápagos, but also because Darwin's finches could not be better ambassadors for evolution. The Grants' research is the subject of the Pulitzer Prize–winning book *The Beak of the Finch* by Jonathan Weiner.

History helped to carry the story of one of my own discoveries about electric eels. In chapter 1, I told you about using my own arm to test out the circuit that forms when an electric eel emerges from the water to shock a threatening animal. I neglected to mention that I discovered this unusual defensive behavior about a year earlier, while working with eels in my laboratory.

I had recently purchased a new net for transferring electric eels from one aquarium to another, and it happened to have a metal rim and handle (I wore rubber gloves, so I was in no danger of getting shocked). The first time I approached a large eel with the net, the eel emerged from the water while pressing its lower jaw against the handle, all the while giving off volleys of high-voltage pulses.

Next I began to investigate the electrical circuit that develops when the eel attacks, but I was also intrigued by a legendary story from the year 1800, told by the renowned naturalist Alexander von Humboldt.[43] Humboldt was on a quest for electric eels on his journey through South America, and a group of fishermen told him they would collect the eels by "fishing with horses." They proceeded to gather a herd of horses and forced them into a shallow pool full of eels. The result was a terrifying spectacle—the eels emerged from the mud and attacked the horses repeatedly, until eventually the eels were exhausted (and two of the horses had drowned). At that point, the eels were safely collected.

Many people thought the story apocryphal—no one else in the last two hundred years had reported electric eels taking the offensive, so to speak, against larger animals. And yet I had just discovered an electric eel behavior that fit perfectly with

FIGURE 4.12. Illustration depicting the battle between eels and horses observed by Alexander von Humboldt in 1800. Notice the electric eel emerging from the water to press its lower jaw against the horse.

Humboldt's tall tale. The missing link was the leaping out of the water part. In his original publication, Humboldt wasn't very clear about the manner of the eels' attack, so I decided to do some more historical research.

That's how I found a remarkable illustration from the 1800s, commissioned by a friend of Alexander von Humboldt, showing the horse-eel battle in great detail (see figure 4.12).[44] In the middle of the image, an eel is shown emerging from the water and pressing its lower jaw against a horse, exactly as I had seen eels behave in my own laboratory. The image, in combination with my findings in the lab, vindicated Humboldt. Needless to say, the addition of this fascinating historical link to the experimental results gave the story much wider appeal. The paper was published with the title "Leaping Eels Electrify Threats, Supporting Humboldt's Account of a Battle with Horses."[45]

Extending the Frontier

Often, perhaps most of the time, scientists are not pursuing a deep mystery or trying to settle a long-simmering debate. Rather, they may be engaged in the less-dramatic work of simply mapping and extending the frontier in their field. This could include taking advantage of new technology to make increasingly more precise measurements of a known phenomenon, or describing the inhabitants of a newly explored ecosystem, or turning a telescope to a new part of the night sky, to give a few examples. This kind of exploratory and descriptive work is key to scientific advancement, though its importance is not always appreciated.

Consider Henrietta Leavitt's work in the early 1900s, cataloging the locations of stars that had periodic variations in brightness—so-called Cepheid variables.[46] In the course of this seemingly routine exercise, Leavitt discovered a relationship between the brightness of a Cepheid variable (star) and its period (how often it completed a cycle of brightening and darkening). The finding became known as Leavitt's Law. It's hard to overstate the importance of the discovery—it gave astronomers a so-called "standard candle" of star brightness.

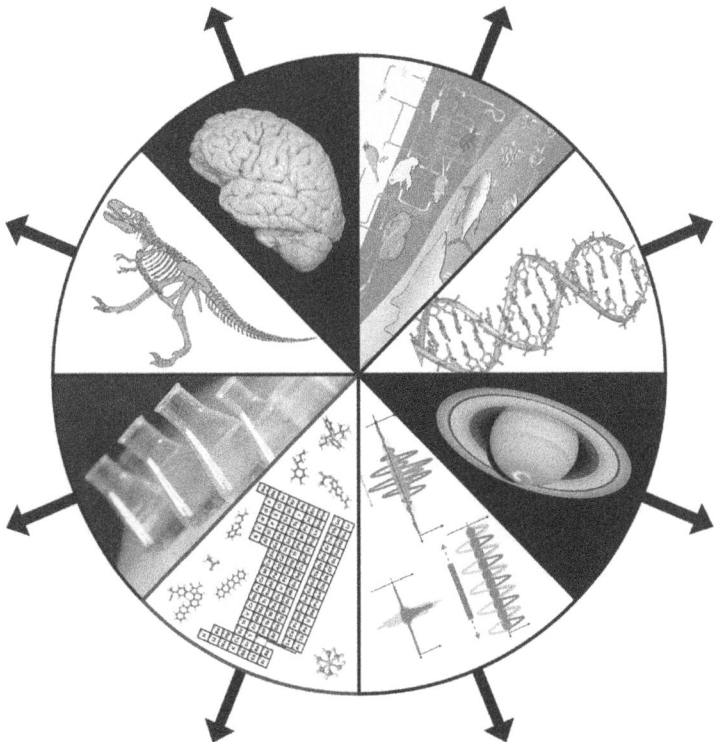

FIGURE 4.13. Although it's not always appreciated, much of science is simply aimed at expanding the frontier of knowledge in a particular field.

All it took next was to determine the distance to some of the Cepheid variables, and astronomers finally had a yardstick with which to measure the size of the visible universe. This followed because once *absolute* brightness of a family of stars was known, differences in their *apparent* brightness could be accurately attributed to distance. Leavitt's discovery was built upon to determine that other galaxies exist, and that they are very, very far away. You might even say that Leavitt's work, ultimately, extended the frontier of astronomy by orders of magnitude.

As for mapping and cataloging, I have been involved in a number of studies that have mapped out the organization of brain areas in diverse species—expanding our understanding of how brains are organized in mammals. That information, in turn, helps us determine how the first mammal brains were organized. The question being, what was the starting point that led, eventually, to our own large brains? It's a big question, but it's not always obvious how seemingly minor pieces of this puzzle can help us find the answer. In such cases, it's best to be explicit.

Here's a quote from the introduction of our paper describing the brain of a humble opossum:

> Our study of how the neocortex of the short-tailed opossum is divided into sensory areas is part of an extensive effort to determine how neocortex is organized in small-brained mammals. A major goal of these studies is to determine which features of cortical organization are common across mammals and which features are unique to specific lineages. By determining how brains have diverged in organization and are specialized in various lines of descent, we may deduce what the brains of the first mammals were like.[47]

In the absence of such an explanation, the research into opossum brains is hard to fit into the bigger picture of neuroscience. Notice we did not simply suggest we were extending the frontier of knowledge in a particular area—that's not enough. Instead we gave a good reason *why* extending this particular frontier was important to answer a much bigger question about brain evolution.

Whimsy and Humor in Science

There seem to be two schools of thought about being light-hearted or funny when describing the results of scientific research. The usual advice is to avoid attempts at humor in published papers. After all, doing science is serious business and the results of research underpin our understanding of the natural world and guide best practices in medicine, engineering, food science, and more. Investigators who make light of their work (so the argument goes) will not be taken seriously, and neither will their results for that matter.

On the other hand, there's growing recognition that scientists need to communicate their findings more broadly and learn how to engage the public whenever possible. Being light-hearted and having a sense of humor makes scientists approachable (so the argument goes), drawing attention to the research and conveying some of the fun of doing science.

As with most things in life, there's probably some truth to both sides of the argument. Winning a Nobel Prize and attending the formal ceremony with the king and queen of Sweden is nearly every scientist's dream. But what about the Ig Nobel Prize for strange and funny research? I should explain that the Ig Nobel Prize is meant to "honor achievements that first make people laugh, and then make them think."[48] Nominees are usually contacted ahead of time and offered the chance to decline the prize and avoid what they might consider embarrassment. Some scientists are delighted to win an Ig Nobel

FIGURE 4.14. The most famous photo of Albert Einstein, taken at his seventy-second birthday party at Princeton University.

Prize, whereas others consider the prize insulting or damaging to their career. It's telling that most winners accept the award and enjoy the irreverent ceremony, often dressing in costumes that commemorate their prize-winning work.[49]

I've brought up the Ig Nobel Prize because the list of winners is a good place to look for publications that have a humorous side. One of my favorite examples is "Neural Correlates of Interspecies Perspective Taking in the Post-mortem Atlantic Salmon: An Argument for Multiple Comparisons Correction."[50] In this study, the investigators examined brain activity in a dead fish presented with pictures of humans in different social situations (using functional magnetic resonance imaging, or fMRI). It sounds utterly ridiculous. But in truth the research was important because the results—purporting to show brain activity in a dead fish—demonstrated how easy it is to get false positives if you don't use the proper statistics. The research has been cited hundreds of times as a cautionary tale, and it certainly meets the criteria of first making you laugh, and then making you think.

Another prize-winning paper was titled "Frictional Coefficient under Banana Skin,"[51] answering the age-old question of slipperiness between a banana peel and the floor. Or consider the epic mystery solved in the study titled "How Do Wombats Make Cubed Poo?"[52]

My own opportunity to pitch the amusing side of a scientific discovery came while studying the parasitoid emerald jewel wasp that is said to make zombies out of cockroaches as food for its young. As a reminder of the story described in chapter 1, the wasp pacifies the American cockroach with a sting to the brain, after which the roach is often called a "zombie," because it can still walk, but doesn't try to escape.

While filming the battle between the wasp and the cockroach in slow motion, I discovered that many of the largest cockroaches successfully defended themselves by kicking the wasp repeatedly in the head with their long, spiny back legs.[53] When that failed, they continued to fight the wasp by raking the wasp with their spiny legs, holding back the stinger with their legs, and (failing that) biting at the wasp's abdomen during the sting.

I had discovered strategies that cockroaches use to avoid becoming a zombie in the first place. It was the perfect setup for an entertaining title and conclusion, so I submitted the paper under the title "How Not to Be Turned into a Zombie." The concluding sentence of the abstract is also an obvious nod to popular culture: "Thus, for a cockroach not to become a zombie, the best strategy is: be vigilant, protect your throat, and strike repeatedly at the head of the attacker." Better yet, the editors at the journal (*Brain, Behavior, and Evolution*) got in on the fun and published the paper on Halloween.[54] As you can imagine, it got a lot of attention from the press. (I later learned of a 2011 Centers for Disease Control blog post titled "Preparedness 101: Zombie Apocalypse"—which got so much attention it crashed the CDC website.)[55]

Having said that, I also want to emphasize the dual nature of the publication. The allusions to popular culture were whimsical to showcase the fun side of doing science. But I'm proud of the science detailed in the paper, which includes slow-motion microscopic interactions between two insects that are both of interest to many biologists studying or teaching about insect brains and behavior. To highlight one important finding—I was able to show that the difference between life and death for the cockroach often comes down to just a few milliseconds in reaction time. This, in turn, helps to explain why neurons that mediate escape are built for exceptional speed (having an extremely large diameter). The latter is often taught in neurobiology courses, but it's hard to find specific examples demonstrating why such speed is needed.

I'm a fan of Halloween and zombie movies, so I had fun with the topic. But I want to end this section on a cautionary note. As a more senior scientist with many publications, I could risk an offbeat title without much concern. I think the risk of not being taken seriously by colleagues is greater at an early career stage (though I have no data on the subject). Plus there's something else to consider—you might succeed with humor beyond your expectations. A hilarious publication that gets a lot of attention could overshadow much of your subsequent research, and that's no laughing matter.

Discovering Stories

Having just described myriad ways you might tell the story of a discovery, I want to dispel a misconception that may have entered your mind. Namely, that you can envision one of these categories ahead of time and use it preemptively as the framework in which to cast any given result. Banish the thought. It's the other way around.

Which is to say, the search for a story should not be approached as a literary gimmick for getting attention or impressing editors (with the obvious exception of whimsy and humor, hence the disclaimers for caution). Putting your results into the right context may indeed impress editors and your audience, but only if the connections that are made, and the revelations that follow, are genuine.

To be sure, some stories come ready-made, such as solving a long-standing mystery or finding definitive evidence supporting one side of a scientific debate. But for the most part, and certainly in my own case, the best stories emerge organically from detective work in the literature, and I hope it is obvious that this could include a range of compelling contexts that are not included above (e.g., vaccine research aimed at saving lives).

Moreover, whether you are conducting research in the laboratory or in the literature, the pitfalls and the strategies for success are remarkably similar—in both cases curiosity and engagement are the key ingredients. Remember the Pulitzer Prize–winning journalist (and science writer) Jon Franklin and his prescient question about scientists—*why it is that some researchers consistently make major discoveries?* His answer is telling (I have paraphrased for gender neutrality):

> The answer is the successful scientist is the one who invests maximum amount of time exploring the branch of nature they're studying—they're out there looking. Beyond that, they actually care, in a human sense, about nature; because they care they keep their eyes open and do their best to understand what they see and fit it into some kind of conceptual framework. . . . Because they're out there looking, the odds are they'll eventually stumble over an important clue of one sort or another.[56]

Franklin was advising budding journalists on "Stalking the True Short Story," and his main point, with which I wholeheartedly agree, is that the same approach that works for making discoveries in science works for discovering stories. You must care enough to be drawn into the subject, pay careful attention to whatever you find, and not assume you know how things will turn out.

I think getting drawn into a subject comes naturally to most people—after all, if you're curious enough to conduct experiments and make a discovery, you're already invested. But, as with doing laboratory research, in addition to motivation, it's useful to have some strategies. Where should you start, and how do you keep from losing your way?

I recommend starting at the beginning. I'm not being flippant; when I do background work to investigate an area of science, I do my best to go back in time as far as possible and read the original papers that kicked off the subject. I like to put myself in the position of those who first started thinking about a topic. This might take me to the 1800s or even earlier. Of course, for many modern areas of research in, say, genetics or physics, the important papers are much more recent. I don't mean to suggest I read everything I find; rather, I use the same approach I take with experiments (or when walking in the woods). I start by trying to get a broad overview, then I check out different leads that grab my attention, only following the most interesting and relevant.

There are many benefits to looking back in time. You will often find that earlier papers were written with the aim of introducing readers to a subject, and sometimes that's just what you need. When I'm out of my element, I work my way forward to the more specialized papers, absorbing the history and background along the way.

Thomas Kuhn weighed in on the chronological trend of scientific publications, suggesting that as scientists become more specialized, their publications will become "oppressive," no longer addressed to anyone who might be interested in the subject.[57] "Instead they will usually appear as brief articles addressed only to professional colleagues . . . whose knowledge of shared paradigm can be assumed and who prove to be the only ones able to read the papers addressed to them."

If you find yourself oppressed in this way, know that it happens to most everyone entering new territory, and consider starting further back in time and then working your way forward again.

And don't dismiss papers just because they are in another language. These days you can copy text from a PDF and paste it into Google Translate to get at least a rough translation of key passages. If that sounds extreme, consider how few people bother to take this extra step. In the process you might well become better read on a subject than the heads of many laboratories.

In my experience, there is almost no research that doesn't have some connections to other scientific subjects you may already know something about or that may touch on the tantalizing "human interest" side of doing science. Let me give an example that's close to home.

I could not help but explore the history of the "panda's thumb" story I described at the beginning of this chapter. In his famous essay on the subject,[58] Gould described his primary source for the anatomical details of the panda's forelimb as the lengthy and detailed morphological study (a so-called monograph) published in 1964 by D. Dwight Davis,[59] who worked at the Chicago Natural History Museum. As Gould suggested, this is an incredibly detailed and impressive work (over three hundred pages in length) covering all aspects of the panda's anatomy. (The details of the forelimb are just a small component, and there was little discussion about how or why the "thumb" evolved.) A bit more sleuthing revealed that Davis spent twenty-five years working on his panda study[60] and published it two months before his untimely death from lung cancer in 1965 (Davis was a smoker).[61]

That got me curious about historical smoking trends in the United States, and I learned that, in 1965, 42 percent of adult Americans smoked, and that the landmark Surgeon General's report associating smoking with cancer had only come out a year earlier, in 1964[62]—much too late for Davis to consider. I also learned that Davis worked with, and was a close friend of, world-renowned herpetologist Karl Schmidt.[63] I had found a connection.

I teach about Karl Schmidt when I cover venoms and the nervous system, because Schmidt famously died after handling a highly venomous snake (an African boomslang).[64] Schmidt thought the snake was safe to handle because it was "rear-fanged," but it nevertheless bit him on the thumb and injected venom. Instead of going straight to the hospital, Schmidt kept a (brief) diary of the venom's effects, dying twenty-four hours later (this story, and Schmidt's diary, were featured on *Science Friday*).[65] Because of such accidental snakebites, these days there are much stricter rules for handling venomous reptiles.

After following that lead, I returned to Davis's panda work and delved deeper. I was surprised to find that Davis, but not Gould, referenced a previous, 1939 essay published in *Nature* titled "The Thumb of the Giant Panda."[66] As it turns out, Gould was not the first to champion the panda's thumb as an unusual contrivance of evolution resulting from historical constraints. In 1939 Frederic Wood Jones published much the same thing.[67] As did Gould, Jones recounted how pandas at the zoo handled bamboo, described the unusual "sixth digit" formed from the radial sesamoid bone of the wrist, and high-lighted this remarkable adaptation as the result of constrained evolution. That previous essay seems lost to history.

Everything I just described is simply the flotsam and jetsam of my own curiosity-driven meanderings through the literature. And yet there was a rich landscape of scientific and historical tidbits for me to pick up, turn over, and consider. As is the case for discoveries in the laboratory, if you're curious, the challenge isn't finding something interesting; the challenge is deciding which interesting thing to pursue.

5

Failure

FIGURE 5.1. The sinking of the *Titanic* by German artist Willy
Stöwer, 1912.

What a wonderful stimulant it would be for the beginner if
his instructor, instead of amazing and dismaying him with
the sublimity of great achievements, would reveal instead
the origin of each scientific discovery, the series of errors

and missteps that preceded it—information that, from
a human perspective, is essential to an accurate
explanation of discovery.

—SANTIAGO RAMÓN Y CAJAL[1]

FAILURE IS JUST A WORD, and it's good to remember that
we give words their meanings. Is Pluto a planet? Are those two
closely related groups of animals separate species? Is your
thumb a finger? It all depends on how we define words and
how we use words. Failure is a good case in point. A common
definition of failure seems to be "lack of success," which is
surely punting as definitions go. So what does it mean to fail?

Clearly some examples of failure have sharp, obvious
borders—such as failing a class, or having a job or school application rejected. Then there are more subtle failures, as in
falling short of your own goals (everyone wants an A in the
class), or perhaps falling short of someone else's expectations.
And don't forget comedic, epic fails (comedic to the beholder
anyway) found on YouTube, as when the new owner of a muscle car tries to show off and ends up skidding into a row of
parked cars. Then there are motivational twists on failure—
like Samuel Beckett's immensely popular quote to "fail better."
That quote is taken out of context from a decidedly unmotivat-
ing piece.[2] That said, there are certainly better and worse ways
to fail. Some failures are unfortunate ends in themselves,
whereas some successes *depend* on previous failures. Failure is

a vast and complicated topic, especially in science. In fact there have been entire books written on failure in science.[3] Given the immense range of meanings and contexts for failure, I should tell you why I decided to include a chapter on the subject, and what I hope to convey.

I have two goals. The first is to provide a confessional of sorts, in the spirit of Cajal's quote above. Which is to say, I hope to counterbalance my history of presenting successes in talks, papers, seminars, and my recent book[4] with a more realistic view of failure rates in the career of at least one scientist. As it turns out, I could give more, and longer, presentations on my failures than on my successes.

I'm not the first to pull back the curtain on personal failures in this way—I was inspired by the concept of "a CV of failures," suggested by Melanie Stefan in her 2010 column in *Nature*.[5] I took up the challenge and searched through my computer files, my paper files, and my memories for failures in school applications, job applications, submissions of papers for publication, grant applications, and finally research projects. I was rewarded for my efforts with a plethora of punishing past experiences, and I'll add that the figures I'll show are certainly an underestimate of my failures. This bias in memory is itself an important point that I'll expand on later.

My brief exposé of failures is primarily meant to reinforce the truism that failures are common in the course of any scientific career, and to give some idea of their scope. Hopefully this brings some solace, and perhaps even acts as Cajal's "wonderful stimulant" for those experiencing their own failures.

My second goal is to explore the kinds of failures that occur in the course of scientific research. Here failure is much more common than would be expected from the published accounts found in peer-reviewed papers. I think it's more common than even investigators themselves usually recall in retrospect. But there is more to say than simply, "Take comfort, failure is common." In the case of research, the ubiquity of failure means that better and worse ways to fail matter—a lot. And that, in turn, means there is advice to be had. Though I will add that, as with other aspects of doing science, each scientist's approach to failure may be different. I'll fall back on Stephen King's disclaimer from his memoir on writing: "This is how it was for me, that's all."

CV of Failures

I've abbreviated my failures into some easily digestible pie charts that give you percentages at a glance (see figure 5.2). What is to be made of these numbers? The most important conclusion is that failure exceeds success in every category. And on some occasions, my career in science lay in the balance between these successes and failures.

I won't tell the roller-coaster story for every ratio shown in the figure. And I'm going to leave out stories about failed job applications, journal submissions, and grant applications—there's not a lot to say other than that failure is part of the process. Instead, I'm going to concentrate on the first chart, showing graduate school applications, and the last chart, showing research ideas. There is a relationship between the two, and some drama hidden behind the figures—at least for me.

My graduate school applications are spread over several years, and I've listed all that I can remember (applications at

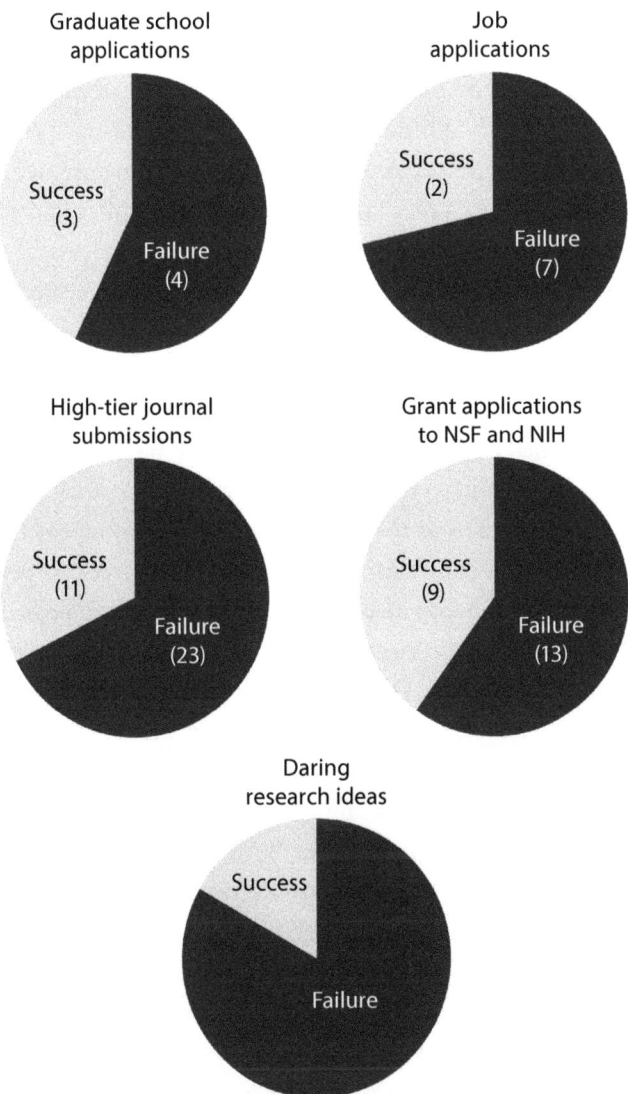

FIGURE 5.2. Pie charts showing successes versus failures for different endeavors. The high-tier journal category includes those with an impact factor above 10.

the time were completed by hand and snail-mailed, so I have no digital record). In 1989 I applied to cell biology programs at Johns Hopkins University and the University of California, Berkeley. But I had little real passion for cell biology, and no doubt this was reflected in my statement of purpose, which, more than thirty years later, I can still remember as being tepid. Needless to say, I was rejected.

In the meantime, I was given special, temporary admission to the graduate program in biology at the University of Maryland, College Park, where I had been an undergraduate. That might seem strange, and perhaps it was. But I got in because I had a sponsor who wanted me to be their graduate teaching assistant, so that I could run their laboratory course and take on much of the test writing and grading for their separate, large lecture course. In exchange I received a stipend, but that's not the only reason I took the temporary position. It also allowed me to continue as a volunteer at the National Zoo in Washington, DC, where I was the lead researcher working on an exciting project focused on star-nosed moles.

As you will see, this project was to be the first data point in my "daring research idea" pie chart. My category of daring ideas (which could equally mean daring hypotheses) is meant to include a specific kind of problem—flights of scientific imagination that generate exciting questions but also tend to be risky propositions. They are often long shots, and answering the question takes time, effort, and resources. On the flip side, such daring ideas can have a big pay-off—if they turn out to be correct. The star-nosed mole project I was working on epitomized this kind of hypothesis (though I should add that the idea for the study did not start with me).

For decades biologists had wondered why star-nosed moles have a star, and what it might possibly be used for. The curator of mammals at the zoo was especially interested in this question, and one day while listening to the radio,[6] he heard a report about the discovery of an "electric sense" in the duck-billed platypus.[7] Scientists in Australia had just shown that the strange duck bill of the platypus is covered with sensors that are used to detect electric fields (so-called electroreceptors) that help the platypus find food (prey animals in water give off electric fields). And with that inspiration, he had a hypothesis: What if the mole's mysterious star was used to detect electric fields? I was a volunteer at the zoo with the job of conducting experiments to test this possibility, and the temporary year as a graduate student at the University of Maryland allowed me to continue the work.

To make my long research story (full of fields trips to Pennsylvania and overnight experiments at the zoo) short, the mole's star is not used to detect electric fields. Although this was my first mark in the failure column, there was an upside to the experience. I learned a lot about an unusual species, and I narrowed my interests to neurosciences and animal behavior. Also, during my year working as a teaching assistant I met a visiting professor of neurosciences (Glenn Northcutt) from the University of California, San Diego, and his lab seemed a great fit for my newfound interests.

I reapplied to graduate school, this time with better focus and a connection to a specific lab, and I was offered admission to UCSD in 1990. As an aside, I also applied to the University of Florida, Gainesville, with an eye toward herpetology, and I was offered admission after an interview (my zoo experience helped). But I wasn't sure whom I would work with, and I was disturbed by crime reports that headlined local newspapers.[8] This was foreshadowing—had I accepted the offer, my first semester would have been during the horrific killing spree of the "Gainesville Ripper," one of the most notorious serial killers in history, who targeted University of Florida students.[9]

In any case, I chose the sunny shores of UCSD, and Glenn Northcutt's lab, which was located practically on the beach at Scripps Institution of Oceanography. It was a welcoming and supportive environment, and things were great for the first year as I took classes and learned new research techniques. But by my second year I hadn't found a specific research project that could hold my interest. So I tried to come up with my own project with my own daring research idea.

By then I had learned a lot about the electric sense in various animals, especially electric fish, which are famous for generating and detecting electric fields as part of a sensory "radar system." A number of scientists wondered if there might be predators that detected electric fish and homed in on their signals for an easy meal (there's an analogy in modern warfare—some missiles home in on enemy radar signals). Such predators *seemed* like something that should have evolved. I thought of the perfect candidate from my work at the National Zoo— the bizarre-looking matamata turtle (*Chelus fimbriata*).

Matamata turtles are fish-eaters that live in the Amazon, in the same habitat as electric fish. The turtle is camouflaged, with a leaf-shaped head and a shell that looks like a piece of wood, and it hunts by waiting patiently, hidden around logs and leaves, often in murky water. This is exactly where you find electric fish. The turtles have tiny eyes, but somehow, they detect nearby fish and catch them with an explosive attack—opening their wide mouth and throat to suck them in. How do they know when a fish is nearby? Especially at night? This seemed like the perfect candidate for an electric fish–detecting predator. The story sounded so good, as I told it to myself, it just had to be true. (My attitude is a good example of what has been called "the planning fallacy," which I'll come back to.)

I found a breeder for these rare turtles in New York State, and my brother (who lived in New York City at the time) picked them up for me. I saw my brother over the holidays, and returned to San Diego with the turtles. It was easy to get electric fish from one of the San Diego labs, and I set about testing the turtles for responses to fish-generated electricity. To make another research story short, the turtles had no interest in electricity. The matamata turtles were strike two in the daring research category.

I will briefly pause my narrative here, to point out an important, shared characteristic for both the star-nosed mole and matamata turtle hypotheses. As exciting as these ideas sounded, each originated not from some bit of data, however small, but rather from imagining what might be going on—independent of nature. I'm all for imagination and inspirational ideas, but (as you have already read) I have found the best ideas pop from at least a kernel of data.

In any case, I took the latest failure in stride and kept at my studies. Soon I decided it would be a good idea to expand my horizons in a different, exciting direction—sharks. One of the scientists at Scripps Institution of Oceanography (Adrianus Kalmijn) was a world expert on shark senses, and I took an opportunity to "rotate" in his laboratory. Laboratory rotations are a standard practice for graduate students as a way to expand their expertise and perhaps find a research focus for their dissertation. Except scientists come from a range of backgrounds and traditions, and they may have very different ideas about what students should do during a rotation. I spent my time swinging a pickaxe in the sun. I was digging ditches for wiring to help with the construction of Kalmijn's electromagnetic research facility on a hillside near Scripps Aquarium. When not digging ditches, I cared for fish in his lab. I went along with the unconventional work because it seemed like a rite of passage, but I never did, or even saw, any research during my rotation. The experience was a different kind of failure, and I was beginning to feel adrift. It didn't help that I was living far from home.

As time passed the situation became dire—I needed to feel a connection to a research project if I was going to put in the years of work it would take to complete a PhD. Glenn was generous in every way, offering a number of different projects from his lab. But the questions were his own, and there's no easy way to transfer the key ingredient—passion for a subject.

Soon I was in crisis. I was depressed and eventually decided I should leave San Diego and move closer to home. In desperation I applied to a different graduate program at Duke University. In my application, I suggested projects working on mammals, trying to leverage my experience and connections from the National Zoo. But I was putting the cart before the horse—most scientists expect a student to take up a problem from their lab, not bring along their own pet project. Needless to say, I was rejected from Duke (adding to the failure side of my balance sheet).

In the end, Glenn came to my rescue. He suggested I try "riding two horses"—meaning he would support me to work on some of my own pet projects, while I also worked on one of the more mainstream research projects from his lab as a form of career insurance. I was still somewhat obsessed with star-nosed moles, plus I held the key to studying them—I could collect them. I was sure there was a good story to discover in their unexplored biology. Studying star-nosed moles would also require me to travel back to the East Coast to collect specimens, giving me a chance to see my family and friends. It solved the homesickness problem and the research problem—in theory.

Except for one thing—what was I going to study about star-nosed moles? I had already suffered a long, drawn-out defeat helping with the electroreception idea at the zoo. If I was going to turn back to this species, where would I even start? Photography, of course. In fact, my long tradition of starting studies with art began with star-nosed moles. I was pushed in that direction by some more advice from photographer John Shaw. In his book on nature photography, Shaw noted that most photographers concentrate on the same few species— elk, deer, and bison. He went on to say, "Small mammals are particularly overlooked. . . . You would do far better to photograph shrews, star-nosed moles, deer mice, kangaroo rats, and red squirrels than working most big-game mammals." I was shocked. Of all the thousands of species of small mammals to choose from, he had called out star-nosed moles as coveted subjects for photography. I was up for the challenge.

As with the jewel wasps, photographing the moles taught me a lot about their behavior, and through a mutual learning experience (the moles got a lot of food rewards for coming to the surface) I was eventually rewarded too—with many unique portraits. And remember my suggestion that there is nothing like starting out of the research gate with an inspiring image to boost your enthusiasm—I certainly needed a boost at the time. With some unique images in hand, I next moved to the scanning electron microscope to get an image of the star. As you have already read, that's what led to the brain discovery described in chapter 3. My graduate school crisis was finally over.

Here you might recall that this is supposed to be a chapter about failure—I just told a story that ended with success. True enough, but I still shudder to think how much I gambled on a science long shot. I had invested so much time, effort, and hope in making my own discovery, I didn't have the fortitude for another failure. I lucked out with my question (though notice the successful idea about the mole's brain sprang from a kernel of data). Plus I had an exceptional mentor who supported me through my crisis. Graduate school is not the time to take such chances.

Consider the pie chart showing the ratios of success to failure when it comes to daring ideas. To come up with this ratio (which is an estimate) I searched back in my files and memories for research projects that didn't work out—and there have been plenty. In fact, as you would expect, over time my success ratio has gone up—I'm better now at guessing which research projects will turn out to be winners. The estimate that I show, of about one in six successes, is where things seem to stand now. Early in my career, it was more like one in ten.

FIGURE 5.3. A star-nosed mole, photographed by the author.

There have been so many failures, it's hard to keep track. I showed the research failure pie chart to my wife, Liz, also a neuroscientist, and she was surprised by the high percentage of failures—even though she was there when many of them happened. I had to remind her, asking, "Remember the time I brought tetrodotoxin-resistant garter snakes to the lab, to study reaction times?"

"Oh yeah," she said. "I forgot about that."

Or how about the knob-tailed gecko project?

"Right," she said, "I forgot about that too."

And so it went. It turns out failures are forgettable—which is a good thing for morale. To give some idea of the scope, in my lab we have studied geckos, ball pythons, alligator snapping turtles, elephant shrews, grasshopper mice, tetrodotoxin-resistant garter snakes, alligators and crocodiles, giant water bugs, tarantulas, water shrews, star-nosed moles, weasels, emerald jewel wasps, emerald tree boas, mole-rats, voles, electric eels, and more. In some cases, an interesting story has developed. But much of the time, the questions posed didn't lead very far.

I'm not the only one to have this experience in science. When a student asked Nobel laureate Linus Pauling how he had so many good ideas, Pauling replied, "Well, you just have lots of ideas and throw away the bad ones."[10]

Inspired by Pauling's comment, in 2017 John Kirwan (an emeritus professor at the University of Bristol, UK) decided to see if this was true in the case of his own, very successful, career.[11] Looking back, Kirwan concluded that only about 25 percent of his research ideas resulted in a publication, and that much of his research effort went into projects that never produced anything.

Or consider the renowned entomologist Thomas Eisner. In his book *For Love of Insects*, Eisner explained his process of discovery less formally: "I already knew as a boy that if I wanted to see things happen—if I wanted to win the revelatory lottery of nature—I had to buy a lot of tickets."[12] Apparently, I'm not alone in having a low batting average.

So how is it that scientists deal with such long odds when it comes to success in research? You have to know that failure is absolutely integral to the process of doing science, and with this knowledge, you have to accommodate failure. That said, I am going to sharply depart from the common cliché heard in Silicon Valley and popular culture—that failing is a good thing, that you should embrace it, and that you should fail big. I don't think you should aim to fail, revel in failure, or try to fail big. Likewise, I would avoid a surgeon who embraces failure, an architect with failed designs, or a financial adviser who invests in failed stocks. Frankly, failure sucks. Failure is to be avoided. Failure in science is inevitable, but the goal is to fail efficiently by making failures small.

Failures Big and Small

I started this section talking about meanings and definitions, and so at this point I must make some distinctions when it comes to ideas and failures. My category of "Daring Research Ideas" is different from the source of ideas described in chapter 1. You may recall that I concluded chapter 1 by saying I get my best ideas straight from the system under study. This is certainly true; in my experience, ideas that spring from data have by far the greatest chance of eventually leading to success. In contrast, my version of daring research ideas (e.g., guessing that matamata turtles might detect electric fish) is much further removed from data, and therefore much more likely to fail in a big way (fail big because exploring the idea doesn't lead anywhere useful).

But failures come in all sizes. Now I want to turn to failures on a different scale—those smaller failures that occur on the way to success. Many of those failures are underappreciated afterthoughts. If the big failures of daring research ideas are easily forgotten, consider how quickly small failures fade from memory. In many cases, it doesn't even make sense to classify them as failures. Let me give examples of such trivial and forgettable failures from everyday life.

When's the last time you worked on a jigsaw puzzle? If you kept at it, you most likely succeeded. But have you ever considered how many times you tried to fit a puzzle piece into the wrong place? Odds are, your failed attempts far exceeded your successes. If you tallied successes versus failures, what would you find?

Or consider a bird-watching hike. Recently I went to Radnor Lake near Nashville and saw three bald eagles (two parents and a fledgling), quite a success. But I probably turned my gaze to thousands of different locations along the way. Were those glances somehow failures to find birds? Was my bird-watching hike towering with failure? Failures can scale. On the very small end of the scale, such failures aren't even acknowledged. I don't think any of us would, or should, classify the iterative process of fitting puzzle pieces into a larger picture, or looking around for birds, as mostly failures punctuated by occasional success. It is often similar in science.

The problem is, many of the "trial and error" failures that lead to success are not included in scientific publications. This creates a misperception of the process for subsequent read-ers—a misperception that is most acute for new investigators who are not "in the know" from hard-won experience. Worse yet, there is an added, prospective, bias that seems to be part of human nature. Psychologist and Nobel laureate Daniel Kahneman calls this the "planning fallacy"—the tendency to be overly optimistic about how things will turn out.[13] He lists many horror stories of overoptimistic planning by individuals, governments, and businesses that regularly underestimate fail-ure rates, costs, and delays for major projects.

The collision of these two perspectives, one a bias from past literature and the other part of human nature, can lead inves-tigators to design overly complex experiments too early in the process, thus hampering innovations along the way. You don't want to start out with the equivalent of a clinical trial. In his 2015 book *Failure: Why Science Is So Successful*,[14] Stuart Fire-stein points out: "Clinical trials are highly organized, have lots of controls, are hypothesis-based, use sophisticated statistics, cost lots of money, are done by people in white lab coats—they have all the trappings of science. Basic science is just pok-ing around in the lab." I'll add that the comparatively recent, worldwide drama of the clinical trials for vaccines from Mod-erna and Pfizer-BioNTech has further cemented this image of how science proceeds.

And yet the real story behind the development of messen-ger RNA (mRNA) vaccines is long and tortuous, and full of many comparatively small failures in the vein described above. To mention just a couple of the now-famous hurdles that had to be cleared, mRNA in its normal configuration was found to induce a massive immune response (not the kind hoped for with successful vaccines). Scientists Katalin Karikó and Drew Weissman, working at the University of Pennsylvania, discovered, through trial and error, that some modifications of the mRNA allowed the molecule to avoid the immune response[15]—paving the way for mRNA vaccines.

Later, a similar problem was found with the lipid coat that was used by Moderna scientists to shepherd the (now-modified) mRNA into cells. Moderna scientist Kerry Benenato spent a year trying over 100 variations of the lipid compounds to optimize the mRNA delivery envelope. She eventually settled on variant 102, and both Benenato and her lipid variant (named SM102) now hold a place of honor in the history of COVID-19 vaccines.[16] But what about variant 34, or 57, or all the others? These minor failures are presumably forgettable. They are misplaced puzzle pieces along the way to success (see figure 5.4).

If such is the usual way of science, the process is akin to an evolutionary tree with many branch points, some very short, and only some leading to success. The successes spawn longer branches with new and refined experiments. As you can see, I have converged on the ideas suggested in chapter 2, only this time I have concentrated more on failure than success. What are some of the practical implications of this view of science?

Consider the branch lengths that lead to successes and failures. These branches could represent time and resources devoted to testing out various ideas or designing different versions of an experiment. If many of these branches lead to failure, their lengths become an important variable in a scientist's productivity. This raises a key question: How much variability can we

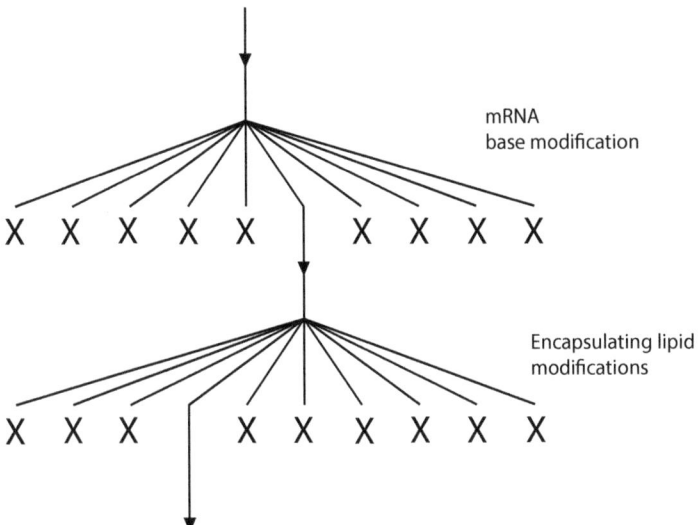

FIGURE 5.4. Schematized view of just a tiny fraction of the steps leading to an mRNA vaccine. This sequence can be contrasted with, for example, vaccine trials that have a more binary structure (either success or failure).

reasonably envision for these branch lengths for similar scientific endeavors? Let me contrast two examples.

Many years ago, entomologist E. O. Wilson used a powerful magnet to test for magnetoreception in ants.[17] It took him only two hours to conclude the experiment was a failure; his ants couldn't have cared less about magnetic fields. Wilson's question was remarkably similar to the question of whether star-nosed moles can detect electric fields. I spent two years working on the star-nosed mole failure. Two hours versus two years. That is easily a thousandfold difference (see figure 5.5). I can't think of a better example of the immensely different

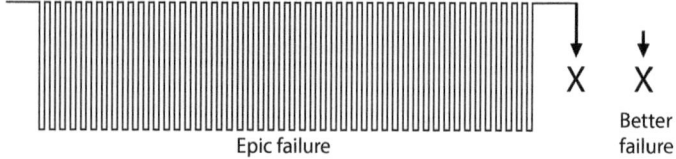

FIGURE 5.5. Two very different paths to failure, one a thousand times longer than the other. A scientist that takes the long route too often is unlikely to get much done.

scales that similar failures can take. In other words, there are better and worse ways to fail.

I'm not the only scientist to learn this lesson the hard way. In her compelling memoir *Lab Girl*, geobiologist Hope Jahren described a summerlong data-gathering failure, and said it taught her the most important thing she knew about science: "that experiments are not about getting the world to do what you want it to do."[18]

Before leaving off this section, I want to return to one of my scientific heroes, Katalin Karikó, who, along with Drew Weissman, pioneered the use of mRNA for vaccine development, leading the way to the COVID-19 vaccines that have saved millions of lives. Karikó famously faced much resistance and had little support in her quest to unlock the potential of mRNA as a therapeutic. So it might be tempting to hold her up as a counterexample—showing that success is actually best achieved by never, ever quitting in the face of constant failure. But that would be grossly inaccurate.

The real story is reflected in Karikó's own words after winning the Nobel Prize: "You don't persevere and repeat and repeat just to say 'I am not giving up.'"[19] Karikó was frustrated not because her research failed, but rather because her pioneering work was met with indifference and lack of funding. Despite being underappreciated at the time, her research results and corresponding publications were major successes that clearly showed her (and later others) the way forward to vaccines and other therapeutics.[20]

Failing Better

One of the unfortunate truisms of life is there's no upper bound to the magnitude and variety of potential failures. So perhaps the first rule for failing better is to avoid the behemoth, epic failures at the upper end of the failure spectrum. What constitutes these in science?

Certainly one peril is falling so in love with a hypothesis that you cannot accept no for an answer. This is sure to lengthen the tortuous path to failure, which must often be walked. It's an all-too-common experience. Nobel laureate Peter Medawar recounted: "Twice in my life I have spent two weary and scientifically profitless years seeking evidence to corroborate dearly loved hypotheses that later turned out to be groundless. . . . It is my recollection of these bad times that accounts for the earnestness of my advice to young scientists that they should have more than one string in their bow and should be willing to take no for an answer if the evidence points that way."[21]

In the worst cases, honest scientists become so enamored with their idea, they begin to unconsciously bias their observations and data and ultimately publish patently wrong results—sometimes even inspiring other labs to follow suit. Famous examples of this unfortunate outcome were highlighted in a colloquium on "Pathological Science" by Irving Langmuir.[22] For example, the case of "N-rays," a mysterious form of radiation that brightened objects in their path. Experiments had to be carried out in very dim rooms, and the effect was near the limits of human perception—a great setup for self-deception. I can recall a children's game: turning down the lights until furniture was barely visible while someone else, usually an older kid, insisted vague outlines were morphing into monsters. It was memorably terrifying. But there were no monsters, and there were no N-rays either, despite the hundreds of papers published on the phenomenon in the early 1900s. (My favorite example: a publication in *The Lancet* proposed that physicians use lack of N-ray emission as proof of death.)[23]

Langmuir's most telling example was of a physics experiment by Bergen Davis from Columbia University involving the unlikely combination of alpha particles and electrons—the so-called Davis-Barnes Effect.[24] The details are not important, so much as Langmuir's experience visiting the lab for a demonstration by Davis, during which it became obvious the effect was a delusion. Langmuir, an expert soon to win a Nobel Prize, pointed out to Davis many obvious problems with the experiments and described how Davis responded: "He immediately—without giving any thought to it—he immediately had an excuse. He had a reason for not paying any attention to any wrong results. It just was built into him. He just had worked that way all along and always would. There is no question but he is honest: He believed these things, absolutely."[25]

Langmuir went on to give some attributes that seemed to characterize pathological science. These included, for example, effects that remain at the limit of detectability; many measurements are necessary because of the low statistical significance of the results, the magnitude of the effect is substantially independent of the intensity of the cause, and criticisms are met with ad hoc excuses. Later, Richard Feynman put the challenge this way: "The first principle is that you must not fool yourself—and you are the easiest person to fool."[26] It's a concise statement of the problem, but short on specific advice. So what can be done?

For starters, the best scientists I know raise their standards in some proportion to their hopes. The more extraordinary the (possible) discovery, the higher the hopes, so a wise scientist follows an innate version of Carl Sagan's famous dictum, "Extraordinary claims require extraordinary evidence."[27] I say wise, because not only is avoiding self-deception best for scientific progress, it's also very much in a scientist's own best interest. In the long run, science has the comforting attribute of self-correction, but it can be hard on the careers of those who get corrected.

Knowing this full well, it has always come as a shock when I've met someone who takes the strategy in the wrong direction—lowering their standards, however unconsciously, in proportion to their hopes. When this happens, there are invariably two related flaws expressed simultaneously. The first is being oblivious to their own bias when it comes to collecting and analyzing data. My favorite comment in this regard is, "John Doe was always the first to notice a trend in the data." Which, in case it's not obvious, is not the compliment it seems (there's often no trend in the data). The second fault is not inviting criticism, and thus becoming oblivious to what their peers think. Maintaining high standards, being open to criticism, and striving for definitive experiments will help you avoid becoming John Doe and following an amazing trend in the data when none is really there.

What else can be done to raise standards of scientific common sense? There may be some small percentage of people who cannot be trained out of an overabundance of creativity (which may be a boon in other walks of life—say, fiction writing). That said, practice is the key to achieving the right balance between hopes and expectations. Perhaps the best kind of practice is getting training on problems that have definitive answers. This follows, because, as outlined at the start of this chapter, most things in science fail. Practice with the definitive, for which failures are obvious, teaches us just how often our ideas are wrong. Hence Niels Bohr's quip that an expert is someone "who has found out by his own painful experience all the mistakes that one can make in a very narrow field."[28] An expert's humbling experience of repeated failure makes them less likely to champion a pet idea over nature's answer.

The Other End of the Stick

It should be obvious that pathological science is, well, pathological. But there are plenty of legitimate and well-reasoned ways to extend the investment in failure. The most understandable of mistakes is to commit to long and complex experiments before all the kinks have been worked out. Working out the kinks is shorthand for finding out, first, whether the whole idea is a failure, and second, whether (once the idea is deemed solid) the experimental design is a failure—or at least not optimal. This is a very real problem for many students and early-career scientists, and it's a natural consequence of the common view of science—that experiments should have all the trappings of a clinical trial. This view, combined with the unrealistic estimate of failure rates, plus the effect of Kahneman's planning fallacy (overoptimism), invites long, drawn-out failures. After all, it's perfectly natural to invest heavily in something you *think* will work. Here again, experience matters.

A particularly telling example of the role of experience (or lack thereof) comes from a report by Neil Glagovich and Arlene Swierczynski from Central Connecticut State University.[29] They tasked graduate students with designing a new laboratory problem for an undergraduate organic chemistry course. The students chose a number of candidate reactions from previously published reports, fully expecting them to work. Instead, most failed—to the astonishment of the graduate students, who "voiced disbelief" about the experience. The students had been taught in classes that demonstrate success. In contrast, the faculty had long experience in the laboratory and hence took frequent failures entirely for granted. There was such an immense gulf between student and faculty expectations that the authors decided to design a new laboratory course demonstrating failure.

Experiments fail for many technical reasons, even when the ideas are good. Perhaps the design of a piece of equipment could be improved, a drug dosage was not optimal, a key variable was not recorded, or maybe controls were inappropriate. Such things are not so much of a problem when you're still feeling things out, but if you have designed a yearlong study involving large groups, mid-course corrections may be impossible.

Plus there are those times when the best mid-course correction is to completely abandon a bad idea. Peter Medawar put this simply, in perhaps my favorite science quote: "An experiment not worth doing is not worth doing well."[30]

With that, perhaps you can imagine a crux for experimental design. I have suggested that success, at least in my own case, depends on refined and definitive experiments plus striving for artistic data. At the same time, I seem to be suggesting that investment in elaborate experiments is a common way to extend failures. How can both be true?

There's no real dichotomy here. Recall from chapter 2 that most experiments evolve over time. Elaborate endpoints garner the attention in final published accounts (e.g., vaccine trials), often obscuring the story of how an experiment evolved from simple beginnings. The process of refinement usually provides a rigorous test of an idea or hypothesis, providing an off-ramp before heavy investments are made. On the other hand, when experiments continue to support an exciting hypothesis, refining them further is one of the great joys of science.

6

Success?

IF THERE'S ONE thing more notoriously hard to define than failure, it's success. Is winning the lottery a success? It *can* be. But it can also strain friendships, eliminate the sense of purpose that accompanies work, and change the dynamic between relatives—not necessarily in a good way. To put a science spin on the problem, the Nobel Prize has brought some scientists practically to tears with a sense of impending doom.[1] That's because the award has often signaled the end of a scientist's ability to "do science." It's not that it takes away opportunity. Rather, as Richard Hamming described the commonly observed effect: "When you are famous it is hard to work on small problems. . . . They try to get the big thing right off. And that isn't the way things go."[2] In other words, winners often feel pressure to work on the most important problems, so they stop working on the small puzzles that are a common gateway to bigger discoveries. As Hamming puts it, they stop planting the little acorns that lead to mighty oaks.

In the same vein, author Elizabeth Gilbert gives an inspiring TED Talk about the personal crisis that descended when one of her books became a bestseller.[3] Much like becoming a Nobel laureate, her success came with immense pressure to create another "big thing." And, as in science, this can overturn the creative process that led to success in the first place. For many people, being able to practice their craft is just as important as what they create.

I'm not shedding tears for Nobel laureates and best-selling authors. These are good problems to have, and when it comes to doing science I think we can agree that Nobel laureates belong squarely in the success category. But these counterintuitive examples illustrate a key point. Even the changes you wish for, and work for, may come with unexpected trade-offs and challenges.

This matters because there are many different ways to be a scientist. You may have been taught that the work of the scientist is putting forward and testing theories, following the scientific method.[4] But you'll never meet *the* scientist doing *the* work using *the* method. That's a good thing, because *the* scientist doesn't sound like a particularly interesting or creative person. *A scientist*, on the other hand, is always a unique person with their own history, their own ideas, their own strengths and weaknesses, and, as a result, their own unique approach to science and career.

Some scientists make doing hands-on science their top priority; others see their most important mission as teaching the next generation; still others get the most out of managing teams as they work together to solve problems. Most often, scientists combine the jobs of doing science, teaching, and managing in various proportions. Although a scientific career comes with many pressures, on the upside there is often leeway for individual scientists to match their own strengths and goals to the needs of an organization or university.

That being the case, it's a good idea to give some thought to what you want, or at least what you think you want, in a career as a scientist. Do you love doing science? Would you like to teach? Can you see yourself managing groups of students as they learn the ropes? What about being part of a team, or running your own team one day, or working in medicine or in industry? Or would you rather do fieldwork in a forest, or chart the heavens from an astronomical observatory? Or some combination of the above?

You might not have the answers to all of these questions, or even most of them. But it's a good idea to think about them regularly. I still think about these questions because the answers can change over time. To give an example, when I was a postdoctoral fellow (a period of research training between getting a PhD and looking for an academic job), I almost swore off a career as an academic entirely, for fear of teaching. I'm a nervous speaker and I couldn't imagine being in front of a class. But I got over my fear, and much to my surprise, I discovered that I love teaching—it's one of my favorite parts of being a scientist, and I wouldn't give it up even if I could.

But there's another question that I didn't think much about: Would I give up doing science to be a manager? I cut my teeth in fairly large labs that were supported by multiple grants, and so I modeled the early years of my career after my mentors' paths. That meant spending a lot of time writing grants, and every hour spent writing a grant was an hour that I couldn't be doing research, or even helping others do research. And the challenge doesn't end when you finally have a grant awarded.

With each grant come progress reports, budget tracking, experimental procedure protocols, safety protocols, compliance inspections, time sheets, and (these days) frequent training to learn new computer interfaces for compiling and submitting all these documents. Of course, none of this includes the work of actually doing research.

Most scientists get informally trained to grow a lab the way you might grow a business. In that case, bigger is almost always better, especially when it comes to the gold standard of evaluation for most universities: funding. But, depending on the discipline, too much success may force you to slowly trade your position as a doer of science to a manager of other people doing science. Hence, as noted by Peter Feibelman: "Given that the job has all these wonderful benefits, you might be surprised that many professors complain about the demands of their work and that many scientists are happy not to be members of the professoriat. . . . Probably the most widespread complaint is that a professor rarely has time to set foot in the lab and to do the scientific research that used to be so much fun."[5]

There's another challenge, too—the pressure to maintain funding often means you must tailor your research toward the mainstream goals of funding agencies—which may, or may not, align with your greatest interests or the most important areas of research. This dynamic can be a very real problem for the advancement of science, as best illustrated (again) by Katalin Karikó's struggle to obtain funding for her mRNA work, which eventually underpinned the COVID-19 vaccines. As she put it, "And as for going where the money was? Well, that I absolutely did not do."[6] She wrote a prodigious number of grants on her mRNA research—in one stretch writing at least one grant a month for two years—but none were ever funded. Luckily for us all, she persevered anyway.

In my own case, I once followed the grant money in the wrong direction. The grant was awarded, and I can recall going out to celebrate—but soon it dawned on me that I was celebrating the ability and commitment to do years of very hard work on topics I wasn't passionate about. As author E. B. White once put it, "No figure is more pitiful to contemplate than a novelist with a thousand-dollar advance from a publishing house and a date when the manuscript is due."[7]

Don't get me wrong, I consider being funded both an honor and a privilege. In fact, to come full circle, it has a lot in common with winning the lottery. For many scientists the lab is a home away from home, and the grants provide the appliances, furnishings, and even the residents. But to do hands-on science on the subjects you love, you might have to settle for smaller grants or find a creative way to support the lab. I have been extremely fortunate (and grateful) to have been funded by the National Science Foundation, which does not focus on health-related topics and hence generally provides smaller grants. I also owe an immense debt of gratitude to the MacArthur Foundation for a five-year, no-strings-attached grant that allowed me to pursue discovery science on topics that were not on the beaten path.

Of course everyone is different. What some scientists avoid, others desire, jumping straight into the fray of the hottest, most competitive, and heavily populated areas of research. I know a very successful faculty member who runs his lab like a CEO, and advocates business classes for his colleagues. All this to say, success means different things to different people, and it may mean different things at different times.

Sometimes, making a change is, paradoxically, how you find out you belong right where you started. Consider Jennifer Doudna, who (along with Emmanuelle Charpentier) won the 2020 Nobel Prize in chemistry for developing the now-famous gene-editing tool CRISPR/Cas9. In 2008 Doudna was disenchanted with academia and left Berkeley to try her hand at the biotechnology company Genentech. But almost as soon as she left Berkeley, she realized the move was a mistake. She preferred being a research scientist in a lab. In her own words, "I didn't have the right skill set or passions to work at a big company."[8] She returned to Berkeley after only two months, and within three years had published the CRISPR work that garnered the Nobel Prize.

You probably won't be surprised to know that, for me, the opportunity to be in the lab or in the field—conducting experiments and observing biology—is the most rewarding part of doing science. I live for those eureka moments when nature reveals something amazing. That's my definition of success, and perhaps the same is true for you. If so, meeting that challenge may require you to think about, and sometimes rethink, your approach, depending on your position and career stage.

I have passed through many phases in my career, having been a research assistant at the National Zoo, a graduate student, a postdoc, a non-tenure-track research assistant professor, a tenure-track assistant professor, and eventually a full professor. During some of that time, I managed a lab supported by large grants and full of students, postdocs, and research assistants, but I've also had a small lab with only a few students, and lately I've flown solo, so to speak, doing science primarily on my own. I have found that, as a rule, each of these career styles has its own glass half-full, half-empty aspect. This is no doubt true for most variations of life in science, so the question then becomes, which parts of the glass are full, which parts are empty, and how does that fit with your own passions?

In a solely research position (much like many medical school faculty) you can often devote most of your time to research without teaching, but you may have to pay your own salary from grants. That means there's more stress in the background when a grant deadline approaches. Typical tenure-track positions provide a hard salary, but in return you must teach courses in addition to conducting research—and this often requires doing less science and more delegating. If you have a large laboratory, you still have the pressure of getting grant money, with the added twist that the careers and livelihoods of other people depend on your success—but with that success may come all kinds of technical and administrative support that can lessen the load. If you chart your course outside of the mainstream, you could end up with less funding and a smaller lab, but paradoxically more freedom—that is, assuming you can do the kind of work that does not require large grants.

This is just a sampling of the many trade-offs and variables that come into play for a single career path. Whole books could be filled (and have been filled) with detailed advice for navigating these waters. For me (always the scientist) I've approached these trade-offs the way some scientists approach a theory. In Hamming's words: "They believe the theory enough to go ahead; they doubt it enough to notice errors and faults so they can step forward and create the new replacement theory."[9] I think that's a good principle not just for science, but for your career as well. You have to start somewhere, so you begin by looking toward mentors as role models. But don't be so sure that you have the same strengths, weaknesses, and passions as everyone else. If you find a new passion along the way, step forward and create something new for yourself.

And speaking of mentors, no chapter on success would be complete without recounting some of my personal experiences in graduate school and as a postdoctoral fellow. I don't feel qualified to say too much in this area, because unlike the diverse studies I have conducted, my career path is only a single experiment. That said, the experiment was a success largely because of my graduate school and postdoctoral mentors, Glenn Northcutt and Jon Kaas, respectively. To me, both are father figures of a sort, having welcomed me into their labs and scientific families. Both gave me steadfast support and encouragement during those precarious early steps as I found my scientific footing. (It would be more accurate to say that science and society in general owe them an immense debt of gratitude for having done much the same for countless other students.) Lucky for me, I don't have any nail-biting stories of triumph over mentorship adversity. That said, I'll highlight a few particulars that seem to offer universal lessons.

I first met Glenn at the University of Maryland when he came to give a seminar. As described in chapter 3, I was working on the (failed) star-nosed mole electroreception project at the time, and because of that, I had become obsessed with the only book on electroreception (at that time).[10] When Glenn sat down with me to talk about my research interests, I pulled out the electroreception book, and talked about gaps in our knowledge and then proposed some experiments. Unbeknownst to me, Glenn was the series editor for that book, so the subject and the research were squarely in his wheelhouse. (I hadn't noticed his name in the small print at the bottom of the cover!) In fact, Glenn studies (among other things) the evolution of electroreception. Which is to say, I *accidentally* did one of the single most important things you can do to impress a potential mentor: I demonstrated at least some understanding and passion for their area of research. That's the lesson: if you are interested in working with someone, learn as much as you can about their research. It's surprising how often students neglect to do this—don't leave it to luck.

One of my favorite early experiences in Glenn's lab was working together on the scanning electron microscope. The machine was a behemoth, with old-fashioned monitors and consoles covered with glowing lights. It reminded me of the bridge of the starship *Enterprise*, which seemed appropriate because Glenn took the role of Captain Kirk, looking at the viewing screen and giving instructions. I was more in the position of Sulu or Chekov, steering the microscope's electron beam to view different areas of each specimen. It was a fun way to collect data, but that was only part of our mission. We were also in search of what Glenn called "killer photos." I learned early on that you can simultaneously collect data and search for inspiring art. It's a lesson that served me well (see chapter 3).

The next lesson was about collaboration. As you read in chapter 3, soon after I used the scanning microscope to examine the nose of the star-nosed mole, I had a question about the mole's brain. Glenn could teach me all I needed to know about sensory receptors in the skin, but we were both out of our element when it came to the neocortex in mammals. Glenn had the obvious answer: find someone who specialized on mammal brains who could help us with that side of the research. Glenn had someone in mind—Jon Kaas at Vanderbilt University. After writing a letter to Jon (this was before email was common) and reading many of his papers for my homework, I drove from the University of California, San Diego, to Vanderbilt University in Nashville, where I was welcomed into another scientific family. That's how, while still in graduate school, I was already collaborating with Jon (who would later become my postdoctoral mentor). The collaboration between the two labs was the perfect recipe for progress. Each lab had a different focus, but once combined, we solved both sides of the equation, so to speak, and soon had a cover article in the science journal *Nature*. The lesson being, if you cannot solve the puzzle on your own, find an expert collaborator.

Later, in Jon's lab, I learned of another collaborative strategy. Namely, most students and postdoctoral fellows worked with other lab members on nearly every project. It was not a competition, but rather a collaboration with one person leading each study. The strategy had many benefits: it gave a feeling of camaraderie among lab members, ensured that each lab member was learning about diverse topics and techniques, and when it came time to publish, multiple authorships were a boost to everyone. This did not detract the least from the leader of the study, who would be the first author. It was a win, win, win strategy that I later adopted when I had my own lab. As a result, students that came through my lab always worked as part of a team and graduated with multiple publications.

Finally, as far as teachers and mentors are concerned, there is one group that is too often overlooked—the technicians and research assistants who form the backbone of many labs and core facilities. These folks are unsung heroes in science and play an important if informal role in teaching and mentoring (it is a foolish student indeed who treats them as "the help"). They often know more about key techniques and equipment than anyone else in the lab, and will take time out of their packed schedules to teach you their secrets. I for one had a master course in scanning electron microscopy and transmission electron microscopy from Charles Graham and Grace Kennedy in San Diego, and much of what I know about sectioning and staining mammal brains I learned from Laura Trice at Vanderbilt University.

Finally, on the darker side, I will add something obvious that should be said nonetheless. Some potential mentors have a well-earned reputation for being unsupportive, treating lab members as indentured servants. Everyone I know who has chosen such a lab, perhaps to learn a new field or get a bump in salary, has paid a price in all the obvious ways—credit for work is held back, there's much conflict and stress, and extracting themselves presents a challenge.

Well, that's a few tidbits of strategy, which I will close by emphasizing one of the greatest assets in science—your passion for a subject. You've probably heard the narrative that many successful people have followed their singular passion all their lives. As for me, my parents recently uncovered one

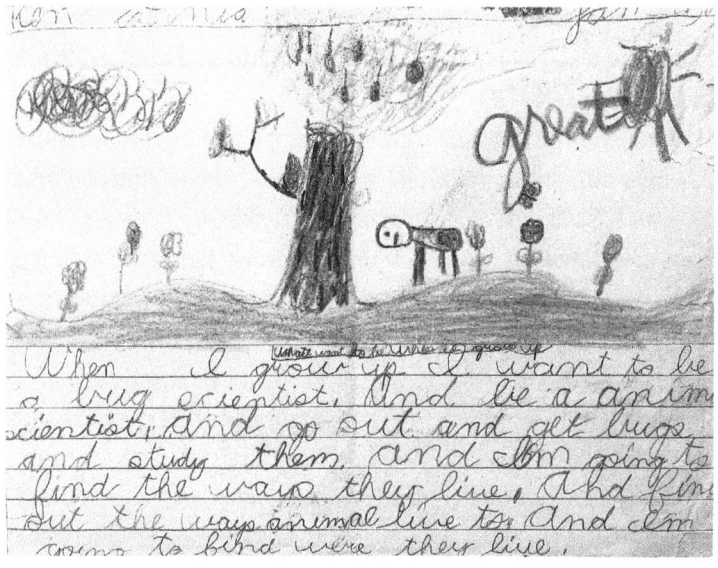

FIGURE 6.1. The author's classroom drawing of "what I want to be when I grow up" (bug scientist).

of my childhood drawings showing what I wanted to be when I grew up (see figure 6.1). I was to be a scientist, studying bugs and animals, and I would find out "the ways they live and where they live." My goal and my stick-figure drawing are eerily reminiscent of the search for the elusive star-nosed mole that formed the basis of my early career. (As an aside, my handwriting and drawing skills have not progressed much from that time, which is to say, you need not be an artist of Cajal's caliber to incorporate art into your science.)

My story appears to follow the script of so many other biologists who seemed to have an innate love of living things from an early age. But what if you didn't grow up with that singular passion for science or nature? What could possibly carry you forward and over all the hurdles?

Not to worry, because here's the thing. Such retrospective reconstructions of the past are not as they seem—they are not fate. If I were a photographer, the script would sound much the same. I would tell you all about my toy cameras and how I pretended to take pictures and record television shows. And how later, my manual Nikon FM camera taught me about aperture, shutter speed, and film speed. Or, if I were a historian, I would be extolling the virtues of Dungeons and Dragons in sparking my imagination, how I was captivated by local Renaissance festivals and living history shows, and how my shop teacher, Mr. Brewer, was an early mentor who encouraged my interest. Or, if I was writing a book, I might talk about Stephen King's memoir on writing. . . .

See what I mean? The same pattern that plays out in scientific discovery works for self-discovery as well. You aren't born with one passion, and you don't have to stick with one passion. If you feed your curiosity with knowledge and experience, something will always grow.

ACKNOWLEDGMENTS

First and foremost, I thank my wife and fellow scientist, Liz Catania, for her help with this book. She is my "ideal reader." She reviewed, and then re-reviewed, every chapter as this project evolved, and her suggestions were always right. Thanks also to Alison Kalett, my editor at Princeton University Press, for believing in the book, helping me work through the early stages, and providing just the right balance of encouragement and critique.

My parents set me on this path of discovery, first by introducing me to nature as soon as I could walk and then encouraging me every step along the way. My brother, Bill, shared many early adventures as we explored the streams, lakes, and beaches of our childhood and later helped me collect star-nosed moles for the National Zoo in Washington, DC. I'm also immensely grateful to my graduate school mentor, Glenn Northcutt, and to my postdoctoral mentor, Jon Kaas. Both welcomed me into their scientific families and gave me the confidence and support to pursue my esoteric interests in unusual creatures.

I am indebted to Vanderbilt University and the Department of Biological Sciences for providing an academic environment where every flavor of research flourishes. Vanderbilt is a great place to do science—and a great place to learn and teach about science. Special thanks to my colleague Maulik Patel for talking science philosophy at our chance meetings during the pandemic—small nudges in trajectory change your destination.

The research that forms the backbone of this work would not have been possible without generous support from several different funding sources. The MacArthur Foundation provided a transformative five-year fellowship that inspired several new lines of research and gave me the freedom to explore new ground. The John Simon Guggenheim Memorial Foundation supported the genesis of my first book, *Great Adaptations*, which was the seed from which the present work arose. Finally, I am grateful to the National Science Foundation, which has steadfastly supported the research in my lab during much of my career (most recently through grant number 2114264).

NOTES

Introduction

1. Zinsser, W. (1995, 4th ed.). *On writing well: An informal guide to writing nonfiction.* HarperCollins; Lamott, A. (1995). *Bird by bird: Some instructions on writing and life.* Anchor; Karr, M. (2015). *The art of memoir.* HarperCollins; Franklin, J. (1987). *Writing for story: Craft secrets of dramatic nonfiction by a two-time Pulitzer Prize winner* (Vol. 2555). Berkeley.

2. King, S. (2000). *On writing: A memoir of the craft.* Simon and Schuster, p. 204.

3. McLeish, T. (2019). *The poetry and music of science: Comparing creativity in science and art.* Oxford University Press.

4. Medawar, P. B. (1996). Is the scientific paper a fraud? In Medawar, P. B. *The strange case of the spotted mouse and other classic essays on science.* Oxford University Press USA, p. 38.

5. Marton, F., Fensham, P., & Chaiklin, S. (1994). A Nobel's eye view of scientific intuition: Discussions with the Nobel prize-winners in physics, chemistry and medicine (1970–86). *International Journal of Science Education, 16*(4), 463.

6. Richter, C. P. (1953). Free research versus design research. *Bulletin of the Atomic Scientists, 9*(9), 323–324.

7. Mehta, G., et al. (2019). Living messages from chemistry icons: Legacies with contemporary relevance. *The Chemical Record, 19*(2–3), 675–686.

8. Hubel, D. H., & Wiesel, T. N. (2004). *Brain and visual perception: The story of a 25-year collaboration.* Oxford University Press, p. 705.

9. Lewis, S. (2016). *Silent sparks: The wondrous world of fireflies.* Princeton University Press. Preface, p. x.

10. McLeish, T. (2019). *The poetry and music of science: Comparing creativity in science and art.* Oxford University Press.

11. King, S. (2000). *On writing: A memoir of the craft.* First Foreword (Oct. 3, 2000). Simon and Schuster.

12. Wilson, E. O. (2013). *Letters to a young scientist.* W. W. Norton & Company, p. 25.

13. Medawar, P. B. (1979). *Advice to a young scientist.* Basic Books.

14. Ramón y Cajal, S. (2004). *Advice for a young investigator.* MIT Press.

Chapter 1: Discoveries

1. Huxley, T. H. (1854). *On the educational value of the natural history sciences.* J. Van Voorst, p. 12.

2. Hubel, D. H., & Wiesel, T. N. (2004). *Brain and visual perception: The story of a 25-year collaboration.* Oxford University Press, p. 705.

3. Parren, S. G. (2013). A twenty-five year study of the wood turtle (*Glyptemys insculpta*) in Vermont: Movements, behavior, injuries, and death. *Herpetological Conservation and Biology, 8*(1), 176–190.

4. Kuhn, T. S., & Hacking, I. (2012). *The structure of scientific revolutions: 50th anniversary edition.* University of Chicago Press.

5. Horgan, J. (2012). What Thomas Kuhn really thought about scientific "truth." *Scientific American* Blogs, Nature Publishing Group, https://www.scientificamerican.com/blog/cross-check/what-thomas-kuhn-really-thought-about-scientific-truth/.

6. Catania, K. (2020). *Great adaptations: Star-nosed moles, electric eels, and other tales of evolution's mysteries solved.* Princeton University Press; Catania, K. C. (2017). Behavioral pieces of neuroethological puzzles. *Journal of Comparative Physiology A, 203*(9), 677–689.

7. Catania, K. C. (2017). Electrical potential of leaping eels. *Brain, Behavior and Evolution, 89*(4), 262–273.

8. Ibid.

9. Kuhn & Hacking (2012). *The structure of scientific revolutions,* p. 36.

10. Catania, K. C. (2017). Power transfer to a human during an electric eel's shocking leap. *Current Biology, 27*(18), 2887–2891.

11. Kuhn & Hacking (2012), *The structure of scientific revolutions,* p. 36.

12. Curie, E. (2001). *Madame Curie* (rev. ed.). Da Capo Press, p. 174. (Original work published in 1938).

13. Kolata, G. (April 8, 2021). Kati Kariko helped shield the world from the coronavirus. *New York Times.*

14. Kuhn & Hacking (2012), *The structure of scientific revolutions*, p. 25.

15. https://www.whitehouse.gov/briefing-room/statements-releases/2022/07/11/remarks-by-president-biden-and-vice-president-harris-in-a-briefing-to-preview-the-first-images-from-the-james-webb-space-telescope/

16. Catania (2020), *Great adaptations*, p. 132.

17. Maurois, A. (1959). *The life of Sir Alexander Fleming.* Jonathan Cape, p. 125.

18. Kuhn & Hacking (2012), *The structure of scientific revolutions*, p. 62.

19. Bruner, J. S., & Postman, L. (1949). On the perception of incongruity: A paradigm. *Journal of Personality, 18*(2), 206–223.

20. Karikó, K. (2023). *Breaking through: My life in science.* Penguin Random House, p. 178.

21. Catania, K. C., Leitch, D. B., & Gauthier, D. (2010). Function of the appendages in tentacled snakes (*Erpeton tentaculatus*). *Journal of Experimental Biology, 213*(3), 359–367.

22. Catania, K. C. (2009). Tentacled snakes turn C-starts to their advantage and predict future prey behavior. *Proceedings of the National Academy of Sciences, 106*(27), 11183–11187.

23. Catania, Leitch, & Gauthier (2010), Function of the appendages in tentacled snakes.

24. Catania, K. C. (2010). Born knowing: Tentacled snakes innately predict future prey behavior. *PLoS One 5*(6), Article e10953.

25. https://www.ted.com/talks/ed_yong_zombie_roaches_and_other_parasite_tales

26. Catania (2020), *Great adaptations*, p. 172.

27. Catania, K. C. (2020). Getting the most out of your zombie: Abdominal sensors and neural manipulations help jewel wasps find the roach's weak spot. *Brain, Behavior and Evolution, 95*(3–4), 181–202.

28. Williams, F. X. (1942). *Ampulex compressa* (Fabr.), a cockroach-hunting wasp introduced from New Caledonia into Hawaii. *Proceedings of the Hawaiian Entomological Society, 11,* 221–233.

29. Catania (2020), Getting the most out of your zombie.

30. Gnatzy, W., Volknandt, W., & Dzwoneck, A. (2018). Egg-laying behavior and morphological and chemical characterization of egg surface and egg attachment glue of the digger wasp *Ampulex compressa* (Hymenoptera, Ampulicidae). *Arthropod structure & development, 47*(1), 74–81.

31. Lamott, A. (1995). *Bird by bird: Some instructions on writing and life.* Anchor, p. 37.

Chapter 2: Experiments

1. Medawar, P. B. (1979). *Advice to a young scientist.* Basic Books, p. 11.

2. Lamott, A. (1995). *Bird by bird: Some instructions on writing and life.* Anchor, p. 21.

3. Zinsser, W. (2006). *On writing well: The classic guide to writing nonfiction.* Harper Perennial, p. 83.

4. Hamming, R. R. (1997). *The art of doing science and engineering: Learning to learn.* CRC Press, p. 9.

5. Hanson, N. R. (1958). The logic of discovery. *The Journal of Philosophy, 55*(25), 1073–1089.

6. Medawar, P. B. (1963). Is the scientific paper a fraud? *The Listener, 70*(12), 377–378.

7. Hanson, N. R. (1958). The logic of discovery. *The Journal of Philosophy, 55*(25), 1073–1089.

8. Skinner, B. F. (1956). A case history in scientific method. *American Psychologist, 11*(5), 221–233.

9. Ibid.

10. Skinner, B. F. (1979). *The shaping of a behaviorist: Part two of an autobiography.* Alfred Knopf, p. 31.

11. Skinner (1956), A case history.

12. Skinner (1979), *The shaping of a behaviorist,* p. 55.

13. Skinner (1956), A case history.

14. Skinner (1979), *The shaping of a behaviorist,* p. 95.

15. Kanwisher, N., McDermott, J., & Chun, M. (1997). The fusiform face area: A module in human extrastriate cortex specialized for face perception. *Journal of Neuroscience, 17*(11), 4302–4311.

16. Tibbetts, E. A., & Dyer, A. G. (2013). Good with faces. *Scientific American, 309*(6), 62–67.

17. Ibid.

18. Ibid.

19. Tibbetts, E. A. (2002). Visual signals of individual identity in the wasp *Polistes fuscatus. Proceedings of the Royal Society of London.* Series B: Biological Sciences, *269*(1499), 1423–1428.

20. Sheehan, M. J., & Tibbetts, E. A. (2011). Specialized face learning is associated with individual recognition in paper wasps. *Science, 334*(6060), 1272–1275.

21. Catania, K. C. (2015). Electric eels use high-voltage to track fast-moving prey. *Nature Communications, 6*(1), 1–6.

22. Ibid.

23. Hanson, N. R. (1958). The logic of discovery. *The Journal of Philosophy, 55*(25), 1073–1089.

24. Medawar, P. B. (1979). *Advice to a young scientist.* Basic Books, p. 16.

25. Zinsser, W. (1995). *On writing well: An informal guide to writing nonfiction* (4th ed.). Introduction. HarperCollins.

Chapter 3: Art

1. Remark made in 1923; recalled by Archibald Henderson, *Durham Morning Herald,* August 21, 1955; Einstein Archive 33–257. And see Einstein, A. (2011). *The ultimate quotable Einstein.* Princeton University Press.

2. Ramón y Cajal, S. (1989). *Recollections of my life.* MIT Press.

3. Cajal (1989), *Recollections,* pp. 40–41.

4. Pannese, E. (1999). The Golgi stain: Invention, diffusion and impact on neurosciences. *Journal of the History of the Neurosciences, 8*(2), 132–140.

5. Cajal (1989), *Recollections,* pp. 307–308.

6. Cajal (1989), *Recollections,* p. 363.

7. Ramón y Cajal, S., & Azoulay, L. (1909). Histologie du système nerveux de l'homme et des vertébrés. In *Histologie du système nerveux de l'homme et des vertébrés.* Maloine, pp. 323–326.

8. Ibid.

9. Catania, K. C., Northcutt, R. G., Kaas, J. H., & Beck, P. D. (1993). Nose stars and brain stripes. *Nature, 364*(6437), 493.

10. Garcia, M. R., & Stark, P. M. (1991). *Eyes on the News.* The Poynter Institute for Media Studies.

11. Catania, K. C. (2007). The evolution of the somatosensory system: Clues from specialized species. In J. H. Kaas & L. A. Krubitzer (Eds.), *Evolution of nervous systems: Vol. 3. Mammals* (pp. 189–206). Elsevier.

12. Catania, K. C. (2001). Early development of a somatosensory fovea: A head start in the cortical space race? *Nature Neuroscience, 4*(4), 353–354.

13. Fry, D. (2012). *Writing your way: Creating a writing process that works for you.* F&W Media, p. 84.

14. Catania, K. C., & Remple, M. S. (2002). Somatosensory cortex dominated by the representation of teeth in the naked mole-rat brain. *Proceedings of the National Academy of Sciences, 99*(8) (April 16), 5692–5697.

15. Hubel, D. H., & Wiesel, T. N. (2004). *Brain and visual perception: The story of a 25-year collaboration.* Oxford University Press, p. 9.

16. Hubel & Wiesel (2004), *Brain and visual perception,* p. 706.

17. Hamming, R. R. (1997). *The art of doing science and engineering: Learning to learn.* CRC Press. Preface (Oct. 28), p. xx.

18. Giaimo, C. (April 1, 2020). The spiky blob seen around the world. *New York Times,* https://www.nytimes.com/2020/04/01/health/coronavirus-illustration -cdc.html.

19. Shaw, J. (1996). *John Shaw's business of nature photography.* Amphoto Books, p. 95.

Chapter 4: Story

1. As quoted in Garner, P. (Nov. 5, 1973). A vanishing breed of film director. *New York Times,* p. 50.

2. Gould, S. J. (2010). *The panda's thumb: More reflections in natural history.* W. W. Norton.

3. Khan, H. A. (2002). The extended panda's thumb and a new global financial architecture, Graduate School of Economics, University of Tokyo, can be downloaded from http://www.eutokyo.ac.jp/cirje/index.htm; Elliott, E. D. (1986). Managerial judging and the evolution of procedure. *University of Chicago Law Review, 53,* 306; Marciano, A., & Khalil, E. L. (2012). Optimization, path dependence and the law: Can judges promote efficiency? *International Review of Law and Economics, 32*(1), 72–82; David, P. A. (1993). Intellectual property institutions and the panda's thumb: Patents, copyrights, and trade secrets in economic theory and history. In *Global dimensions of intellectual property rights in science and technology* (pp. 19, 29). National Research Council/US National Academy of Sciences.

4. Conant, G. C., & Wolfe, K. H. (2008). Turning a hobby into a job: How duplicated genes find new functions. *Nature Reviews Genetics, 9*(12), 938–950.

5. Franklin, J. (1987). *Writing for story: Craft secrets of dramatic nonfiction by a two-time Pulitzer Prize winner* (Vol. 2555). Berkeley. P. 70.

6. Letter from Darwin to Asa Gray, April 3, 1860. See Darwin Correspondence Project: https://www.darwinproject.ac.uk/letter/DCP-LETT-2743.xml.

7. Arditti, J., Elliott, J., Kitching, I. J., & Wasserthal, L. T. (2012). "Good Heavens what insect can suck it"–Charles Darwin, *Angraecum sesquipedale* and *Xanthopan morganii praedicta. Botanical Journal of the Linnean Society, 169*(3), 403–432.

8. Ibid.

9. Ibid.

10. Catania, K. C., Lyon, D. C., Mock, O. B., & Kaas, J. H. (1999). Cortical organization in shrews: Evidence from five species. *Journal of Comparative Neurology, 410*(1), 55–72.

11. Catania, K. C., Northcutt, R. G., & Kaas, J. H. (1999). The development of a biological novelty: A different way to make appendages as revealed in the snout of the star-nosed mole *Condylura cristata*. *Journal of Experimental Biology, 202*(20), 2719–2726.

12. Gould, S. J. (1985). *Ontogeny and phylogeny*. Harvard University Press.

13. Catania, K. C., & Remple, F. E. (2005). Asymptotic prey profitability drives star-nosed moles to the foraging speed limit. *Nature, 433*(7025), 519–522.

14. Krebs, J. R., & Stephens, D. W. (1986). *Foraging theory*. Princeton University Press.

15. Catania & Remple (2005), Asymptotic prey profitability.

16. Griffin, D. R. (1958). *Listening in the dark: The acoustic orientation of bats and men*. Yale University Press.

17. Haskell, P. T., & Belton, P. (1956). Electrical responses of certain lepidopterous tympanal organs. *Nature, 177*(4499), 139–140.

18. Conner, W. E., & Corcoran, A. J. (2012). Sound strategies: The 65-million-year-old battle between bats and insects. *Annual Review of Entomology, 57*(1), 21–39.

19. Griffin, D. R. (2001). Return to the magic well: Echolocation behavior of bats and responses of insect prey. *BioScience, 51*(7), 555–556.

20. Barchan, D., Kachalsky, S., Neumann, D., Vogel, Z., Ovadia, M., Kochva, E., & Fuchs, S. (1992). How the mongoose can fight the snake: The binding site of the mongoose acetylcholine receptor. *Proceedings of the National Academy of Sciences, 89*(16), 7717–7721.

21. Catania, K. C. (2009). Tentacled snakes turn C-starts to their advantage and predict future prey behavior. *Proceedings of the National Academy of Sciences, 106*(27), 11183–11187.

22. Wilson, R. P., Vargas, F. H., Steinfurth, A., Riordan, P., Ropert-Coudert, Y., & Macdonald, D. W. (2008). What grounds some birds for life? Movement and diving in the sexually dimorphic Galápagos cormorant. *Ecological Monographs, 78*(4), 633–652.

23. Burga, A., Wang, W., Ben-David, E., Wolf, P. C., Ramey, A. M., Verdugo, C., Lyons, K., Parker, P. G., & Kruglyak, L. (2017). A genetic signature of the evolution of loss of flight in the Galapagos cormorant. *Science, 356*(6341), eaal3345. PMID: 28572335; PMCID: PMC5567675. https://doi.org/10.1126/science.aal3345.

24. Ratcliffe, J. M., Fenton, M. B., & Galef, B. G., Jr. (2003). An exception to the rule: Common vampire bats do not learn taste aversions. *Animal Behaviour, 65*(2), 385–389.

25. Catania, K. C. (2006). Underwater "sniffing" by semi-aquatic mammals. *Nature, 444*(7122), 1024–1025.

26. Howell, A. B. (1970). *Aquatic mammals: Their adaptations to life in the water.* Dover Books on the Biological Sciences, p. 74. (Original work published 1930)

27. Catania (2006), Underwater "sniffing."

28. Alvarez, L. W., Alvarez, W., Asaro, F., & Michel, H. V. (1980). Extraterrestrial cause for the Cretaceous-Tertiary extinction. *Science, 208*(4448), 1095–1108.

29. Alvarez, L. W. (1983). Experimental evidence that an asteroid impact led to the extinction of many species 65 million years ago. *Proceedings of the National Academy of Sciences, 80*(2), 627–642, at p. 632.

30. Svensson, P. (2021). *The book of eels: Our enduring fascination with the most mysterious creature in the natural world.* Ecco.

31. Schmidt, J. (1923). IV.—The breeding places of the eel. *Philosophical Transactions of the Royal Society of London. Series B, Containing Papers of a Biological Character, 211*(382–390), 179–208, at p. 181.

32. Darwin, C. (1881). *The formation of vegetable mould, through the action of worms: With observations on their habits.* J. Murray.

33. Catania, K. C. (2008). Worm grunting, fiddling, and charming—humans unknowingly mimic a predator to harvest bait. *PLoS One, 3*(10), Article e3472.

34. Valenstein, E. S. (2005). The war of the soups and the sparks. In Valenstein, E. S., *The War of the Soups and the Sparks.* Columbia University Press.

35. Ibid.

36. Loewi, O. (1921). Über humorale Übertragbarkeit der Herznervenwirkung. *Pflüger's Archiv für die gesamte Physiologie des Menschen und der Tiere, 189*(1), 239–242.

37. Valenstein (2005), The war of the soups and the sparks.

38. Catania, K. C., & Kaas, J. H. (1995). Organization of the somatosensory cortex of the star-nosed mole. *Journal of Comparative Neurology, 351*(4), 549–567.

39. Catania, K. (1995). Magnified cortex in star-nosed moles. *Nature, 375*(6531), 453–454.

40. Ibid.

41. Arditti, Elliott, Kitching, & Wasserthal, (2012), "Good Heavens what insect can suck it."

42. Wiener, J. (1995). *The beak of the finch.* Vintage.

43. Humboldt, A. von. (1807). Jagd und Kampf der electrischen Aale mit Pferden: Aus den Reiseberichten des Hrn. Freiherrn Alexander v. Humboldt. *Gilberts Annalen der Physik 25,* 34–43.

44. Schomburgk, R. H. (1843). *Ichthyology: Fishes of British Guiana, Part II* (W. Jardine, Ed.). The Naturalist's Library, Vol XL.

45. Catania, K. C. (2016). Leaping eels electrify threats, supporting Humboldt's account of a battle with horses. *Proceedings of the National Academy of Sciences, 113*(25), 6979–6984.

46. Leavitt, H. S. (1908). 1777 variables in the Magellanic Clouds. *Annals of Harvard College Observatory, 60,* 87–108.

47. Catania, K. C., Jain, N., Franca, J. G, Volchan, E., & Kaas, J. H. (2000). The organization of somatosensory cortex in the short-tailed opossum (*Monodelphis domestica*). *Somatosensory & Motor Research, 17*(1), 39–51.

48. https://improbable.com/ig/about-the-ig-nobel-prizes/

49. Ibid.

50. Bennett, C. M., Miller, M. B., & Wolford, G. L. (2009). Neural correlates of interspecies perspective taking in the post-mortem Atlantic Salmon: An argument for multiple comparisons correction. *NeuroImage, 47*(Supplement 1), S125.

51. Mabuchi, K., Tanaka, K., Uchijima, D., & Sakai, R. (2012). Frictional coefficient under banana skin. *Tribology Online, 7*(3), 147–151.

52. Yang, P. J., Chan, M., Carver, S., & Hu, D. L. (2018). How do wombats make cubed poo? In *71st Annual Meeting of the APS Division of Fluid Dynamics, 63.*

53. Catania, K. C. (2018). How not to be turned into a zombie. *Brain, Behavior and Evolution, 92*(1–2), 32–46.

54. Ibid.

55. Kruvand, M., & Silver, M. (2013). Zombies gone viral: How a fictional zombie invasion helped CDC promote emergency preparedness. *Case Studies in Strategic Communication, 2*(1), 34–60.

56. Franklin (1987), *Writing for story,* p. 70.

57. Kuhn, T. S., & Hacking, I. (2012). *The Structure of Scientific Revolutions: 50th Anniversary Edition.* Chicago: University of Chicago Press, p. 20.

58. Gould (2010), *The Panda's Thumb.*

59. Davis, D. D. (1964). *Fieldiana: Zoology Memoirs: Vol. 3. The giant panda: A morphological study of evolutionary mechanisms.* Chicago Natural History Museum.

60. Moore, J. C. (1965). D. Dwight Davis: 30 December 1908–6 February 1965. *Journal of Mammalogy, 46,* 371–372.

61. https://www.encyclopedia.com/science/dictionaries-thesauruses-pictures-and-press-releases/davis-ddelbert-dwight

62. United States Surgeon General's Advisory Committee on Smoking. (1964). *Smoking and health: Report of the Advisory Committee to the Surgeon General of the public health service* (No. 1103). US Department of Health, Education, and Welfare, Public Health Service.

63. https://www.encyclopedia.com/science/dictionaries-thesauruses-pictures-and-press-releases/davis-ddelbert-dwight

64. Emerson, A. E. (1958). K. P. Schmidt—herpetologist, ecologist, zoogeographer. *Science, 127*(3307), 1162–1163.

65. https://www.sciencefriday.com/segments/diary-of-a-snake-bite-death/

66. Jones, F. W. (1939). The thumb of the giant panda. *Nature, 143*(3613), 157.

67. Ibid.

Chapter 5: Failure

1. Ramón y Cajal, S. (2004). *Advice for a young investigator*. MIT Press, p. 10.

2. Samuel, B. (1983). *Worstward ho*. John Calder.

3. Firestein, S. (2015). *Failure: Why science is so successful*. Oxford University Press.

4. Catania, K. (2020). *Great adaptations: Star-nosed moles, electric eels, and other tales of evolution's mysteries solved*. Princeton University Press.

5. Stefan, M. (2010). A CV of failures. *Nature, 468* (November 17), 467.

6. Zimmer, C. (1993). The electric mole. *Discover Magazine*, August 1.

7. Scheich, H., Langner, G., Tidemann, C., Coles, R. B., & Guppy, A. (1986). Electroreception and electrolocation in platypus. *Nature, 319*(6052), 401–402.

8. The Sentinel Staff. (January 17, 1990). Fotopoulos video shows murder police also find guns, ammunition: [2 STAR EDITION]. *Orlando Sentinel*.

9. Ryzuk, M. S. (1994). *The Gainesville Ripper*. Dutton Books.

10. http://scarc.library.oregonstate.edu/coll/pauling/bond/audio/1977v.66-ideas.html

11. Kirwan, J. (2017). It's good to have lots of bad ideas. *Nature, 548*, 491.

12. Eisner, T. (2005). *For love of insects*. Harvard University Press, p. 395.

13. Kahneman, D. (2011). *Thinking, fast and slow*. Macmillan, p. 249.

14. Firestein (2015), *Failure*, p. 147.

15. Zuckerman, G. (2021) *A shot to save the world: The remarkable race and ground-breaking science behind the Covid-19 vaccines*. Penguin UK, p. 79.

16. Zuckerman, (2021), *A shot to save the world*, p. 154.

17. Wilson, E. O. (2013). *Letters to a young scientist*. W. W. Norton & Company, p. 84.

18. Jahren, H. (2017). *Lab girl*. HarperCollins, p. 75.

19. https://www.nytimes.com/2023/10/02/health/nobel-prize-medicine.html

20. Ibid.; Karikó, K., Buckstein, M., Ni, H., & Weissman, D. (2005). Suppression of RNA recognition by Toll-like receptors: The impact of nucleoside modification and the evolutionary origin of RNA. *Immunity, 23*(2), 165–175.

21. Medawar, P. B. (1979). *Advice to a young scientist.* Basic Books, p. 6.

22. Langmuir, I. (1989). Pathological science. *Research-Technology Management, 32*(5), 11–17.

23. Munro, J. (1904). The "N" rays as a proof of death. *The Lancet, 163*(4207), 1082.

24. Davis, B., & Barnes, A. H. (1929). The capture of electrons by swiftly moving alpha-particles. *Physical Review, 34*(1), 152.

25. Langmuir, I. (1989) Pathological science. *Research-Technology Management 32*(5), 11–17.

26. Feynman, R. (1974). *Cargo cult science: Some remarks on science, pseudoscience, and learning how to not fool yourself* [Caltech's 1974 commencement address].

27. Sagan, C. (1979). *Broca's brain: Reflections on the romance of science.* Random House, p. 62.

28. Coughlan, R. (September 6, 1954). Dr. Edward Teller's magnificent obsession. *Life*, 61–74.

29. Glagovich, N. M., & Swierczynski, A. M. (2004). Teaching failure in the laboratory: Turning mistakes into learning opportunities. *Journal of College Science Teaching, 33*(6), 45–47.

30. Medawar, P. B. (1984). *The limits of science.* Harper & Row, p. 29.

Chapter 6: Success?

1. Hamming, R. (March 7, 1986). *You and your research* [Seminar presentation]. Bell Communications Research Colloquium Seminar, Morristown, NJ, United States.

2. Ibid.

3. https://www.ted.com/talks/elizabeth_gilbert_your_elusive_creative_genius

4. Popper, K. (2005). *The logic of scientific discovery.* Routledge.

5. Feibelman, P. J. (2011). *A PhD is not enough! A guide to survival in science.* Basic Books, p. 73.

6. Karikó, K. (2023). *Breaking through: My life in science.* Penguin Random House, p. 180.

7. White, E. B. (2014). *Writings from "The New Yorker" 1927–1976*. HarperCollins.

8. Isaacson, W. (2021). *The code breaker: Jennifer Doudna, gene editing, and the future of the human race*. Simon and Schuster, p. 201.

9. Hamming (March 1986), You and your research.

10. Bullock, T. H., & Heiligenberg, W. (Eds.). (1986). *Electroreception*. Wiley Series in Neurobiology.

ILLUSTRATION CREDITS

Introduction

0.1. By Charles Catania.

Chapter 1

1.1. By the author as first published in Catania, K. C. (2017). Power transfer to a human during an electric eel's shocking leap. *Current Biology*, 27(18), 2887–2891.

1.2. By the author.

1.3. By the author.

1.4. By the author.

1.5. By the author.

1.6. By the author.

1.7. By the author as first published in Catania, K. C. (2020). Getting the most out of your zombie: Abdominal sensors and neural manipulations help jewel wasps find the roach's weak spot. *Brain Behavior and Evolution*, 95(3–4), 181–202.

1.8. By the author.

1.9. By the author.

1.10. By the author as first published in Catania (2020), Getting the most out of your zombie.

1.11. By the author.

Chapter 2

2.1. Redrawn by the author from images in Skinner, B. F. (1956). A case history in scientific method. *American Psychologist, 11*(5), 221.

2.2. Redrawn by the author from images in Skinner (1956), A case history, 221.

2.3. Redrawn by the author from images in Skinner (1956), A case history, 221.

2.4. By the author, including images from figures 2.1 to 2.3.

2.5. By the author.

2.6. By the author as first published in Catania, K. C. (2015). Electric eels use high-voltage to track fast-moving prey. *Nature Communications, 6*(1), 8638. https://doi.org/10.1038/ncomms9638.

2.7. By the author.

2.8. By the author.

2.9. By the author.

2.10. By the author.

2.11. By the author.

Chapter 3

3.1. By Santiago Ramón y Cajal.

3.2. By Santiago Ramón y Cajal.

3.3. By the author.

3.4. By the author.

3.5. By the author.

3.6. By the author.

3.7. By the author.

3.8. By the author. Upper left image by artist Lana Finch, as first published in Catania, K. C., & Remple, M. S. (2002). Somatosensory cortex dominated by the representation of teeth in the naked mole-rat brain. *Proceedings of the National Academy of Sciences, 99*(8), 5692–5697.

3.9. By the author.

3.10. By Alissa Eckert and Dan Higgins at the Centers for Disease Control and Prevention.

3.11. All images except for crocodile by the author. Crocodile image by Duncan Leitch. Original publication source as follows for each image proceeding from left to right, starting with the top row: Catania, K. C., Northcutt, R. G., & Kaas, J. H. (1999). The development of a biological novelty: A different way to make appendages as revealed in the snout of the star-nosed mole *Condylura cristata*. *Journal of Experimental Biology, 202*(20), 2719–2726; Catania, K. C., & Kaas, J. H. (1996). The unusual nose and brain of the star-nosed mole. *BioScience, 46*, 578–586; Catania, K. C., Leitch, D. B., & Gauthier, D. (2010). Function of the appendages in tentacled snakes (*Erpeton tentaculatus*). *Journal of Experimental Biology, 213*(3), 359–367; Catania, K. C. (2001). Early development of a somatosensory fovea: A head start in the cortical space race? *Nature Neuroscience, 4*(4), 353–354; Catania, K. C. (2000). Epidermal sensory organs of moles, shrew moles, and desmans: A study of the family talpidae with comments on the function and evolution of Eimer's organ. *Brain Behavior and Evolution, 56*(3), 146–174; Leitch, D. B., & Catania, K. C. (2012). Structure, innervation and response properties of integumentary sensory organs in crocodilians. *Journal of Experimental Biology, 215*(23), 4217–4230; Catania, K. C. (2008). No taming the shrew. *Natural History Magazine, 117*, 56–60. From *Natural History*, March 2008, copyright © Natural History Magazine, Inc., 2008; Catania, K. C., Northcutt, R. G. Kaas, J. H., & Beck, P. D. (1993). Nose stars and brain stripes. *Nature, 364* (6437), 493; Catania, K. (2014). The shocking predatory strike of the electric eel. *Science, 346*, 1231–1234.

Chapter 4

4.1. By the author.

4.2. By Montasim Jawar, published in Wikimedia Commons under Creative Commons Attribution-Share Alike 4.0 International.

4.3. Left side plate drawn by the author, modeled from a public domain image in Bailey, L. H. (Liberty Hyde), 1858–1954, ed. Miller, Wilhelm. (1901). *Cyclopedia of American horticulture, comprising suggestions for cultivation of horticultural plants, descriptions of the species of fruits, vegetables, flowers, and ornamental plants sold in the United States and Canada, together with geographical and biographical sketches.* New York, Doubleday, Page & Company. Moth (*right side*) by Esculapio, Natural History Museum of London, published in Wikimedia Commons under Creative Commons Attribution-Share Alike 3.0.

4.4. Photo of certificate, reproduced with permission from Guinness World Records.

4.5. Image by the author.

4.6. Bat drawn by Lana Finch, remainder by the author.

4.7. Left side by Shino Jacob Koottanad, published on Wikimedia Commons and licensed under the Creative Commons Attribution-Share Alike 4.0 International. Right side by David Stanley from Nanaimo, Canada, published under CC BY 2.0 Attribution 2.0 Generic.

4.8. As published in Catania, K. C. (2006). Olfaction: Underwater "sniffing" by semi-aquatic mammals. *Nature, 444*(7122), 1024.

4.9. Photo by the author.

4.10. By the author.

4.11. By the author.

4.12. From Schomburgk, R. H. (1843). *Ichthyology: Fishes of British Guiana, Part II* (W. Jardine, Ed.). The Naturalist's Library, Vol. XL. London.

4.13. Assembled by the author using the following images, moving clockwise from the human brain. Human brain, by the author; food web by LadyofHats, Wikimedia Commons Creative Commons CC 1.0 Universal Public Domain Dedication; DNA by Jerome Walker, Wikimedia Commons released to public domain; Saturn, Hubble Space Telescope, NASA, public domain; wave functions, Maschen, Wikimedia Commons

released to public domain; periodic table and molecules,
László Németh, Wikimedia Commons released to public
domain and Jynto using Discovery Studio Visualizer,
Wikimedia Commons released to public domain; beakers,
PICRYL public domain image of Chemistry-3533039 960 720;
T-rex, Conty, Wikimedia Commons released to public domain.

4.14. Albert Einstein Archives, The Hebrew University of Jerusalem.

Chapter 5

5.1. By Willy Stöwer, 1912. In the public domain.

5.2. By the author.

5.3. By the author.

5.4. By the author.

5.5. By the author.

Chapter 6

6.1. By the author.

INDEX

A NOTE ON THE TYPE

This book has been composed in Arno, an Old-style serif typeface in the
classic Venetian tradition, designed by Robert Slimbach at Adobe.

GPSR Authorized Representative: Easy Access System Europe - Mustamäe tee
50, 10621 Tallinn, Estonia, gpsr.requests@easproject.com